W9-CKI-665

HOW TO PREVENT SPILLS OF HAZARDOUS SUBSTANCES

HOW TO PREVENT SPILLS OF HAZARDOUS SUBSTANCES

by

W. Unterberg, R.W. Melvold,
K.S. Roos, P.A. Scofield

Combustion Engineering
Environmental Monitoring and Services, Inc.
Newbury Park, California

NOYES DATA CORPORATION
Park Ridge, New Jersey, U.S.A.

Library of Congress Catalog Card Number 88-17017
ISBN: 0-8155-1177-9
ISSN: 0090-516X
Printed in the United States

Published in the United States of America by
Noyes Data Corporation
Mill Road, Park Ridge, New Jersey 07656

10 9 8 7 6 5 4 3 2 1

Library of Congress Cataloging-in-Publication Data

How to prevent spills of hazardous substances / W. Unterberg . . . [et al.].
p. cm. -- (Pollution technology review, ISSN 0090-516X ; no. 156)
Bibliography: p.
Includes index.
ISBN 0-8155-1177-9 :
1. Hazardous substances--Safety measures--Handbooks, manuals, etc.
I. Unterberg, Walter. II. Series.
T55.3.H3H68 1988
604.7--dc 19 88-17017
CIP

Foreword

This book is designed to provide technical guidance to prevent spills or releases of hazardous substances from fixed facilities that produce hazardous substances, store them, or transfer them to and from transportation terminals. The audience to be addressed includes managerial and supervisory personnel as well as "hands on" personnel associated with smaller-sized chemical manufacturing facilities. The hazardous substances in question number almost 700, excluding oil, and are those designated pursuant to Section 101 (14) of the Comprehensive Environmental Response, Compensation and Liability Act of 1980, otherwise known as CERCLA or Superfund (Public Law 96-510).

The earlier Clean Water Act (Public Law 92-500) in Section 311, required that the President issue regulations "establishing procedures, methods and equipment. . . to prevent discharges of oil and hazardous substances from vessels and from onshore and offshore facilities. . ." Under a 1973 executive order, the U.S. Environmental Protection Agency (EPA) was to promulgate regulations for preventing nontransportation-related spills.

In 1973, EPA issued oil pollution prevention regulations that require certain fixed facilities to have on file a spill prevention, control and countermeasures plan (SPCC plan). These regulations have been largely accepted by industry and have proven to be effective in preventing oil spills.

In the case of hazardous substances, no similar federal regulations existed at the time this document was written. In its own interest, a large segment of the industry producing, storing, and handling hazardous substances has instituted internal spill prevention plans. However, many of the smaller affected facilities may not have generated spill prevention plans for various reasons. This manual is designed to assist them in preparing such plans.

The information in the book is from *Manual for Preventing Spills of Hazardous Substances at Fixed Facilities,* prepared by W. Unterberg, R.W. Melvold, K.S. Roos, and P.A. Scofield of Combustion Engineering, Environmental Monitoring and Services, Inc. for the U.S. Environmental Protection Agency, issued August 1987.

The table of contents is organized in such a way as to serve as a subject index and provides easy access to the information contained in the book.

Advanced composition and production methods developed by Noyes Data Corporation are employed to bring this durably bound book to you in a minimum of time. Special techniques are used to close the gap between "manuscript" and "completed book." In order to keep the price of the book to a reasonable level, it has been partially reproduced by photo-offset directly from the original report and the cost saving passed on to the reader. Due to this method of publishing, certain portions of the book may be less legible than desired.

ACKNOWLEDGMENT

The preparation of this manual was guided by Leo T. McCarthy, Jr., Project Officer, U.S. Environmental Protection Agency (EPA). His suggestions, assistance and support are hereby gratefully acknowledged.

NOTICE

The materials in this book were prepared as accounts of work sponsored by the U.S. Environmental Protection Agency. This information has been reviewed in accordance with the U.S. Environmental Protection Agency's review policies and approved for publication. On this basis the Publisher assumes no responsibility nor liability for errors or any consequences arising from the use of the information contained herein.

Mention of trade names or commercial products does not constitute endorsement or recommendation for use by the Agency or the Publisher. Final determination of the suitability of any information or product for use contemplated by any user, and the manner of that use, is the sole responsibility of the user. The book is intended for information and guidance purposes only. The reader is warned that caution must always be exercised when dealing with hazardous substances; and expert advice should be obtained before implementation is considered.

Contents and Subject Index

1. Introduction

The purpose of this manual is to provide guidance to prevent spills of hazardous substances from fixed facilities that produce hazardous substances from raw or starter materials as products, byproducts or waste products; store hazardous substances; or transport hazardous substances. The audience to be addressed includes managerial and supervisory personnel as well as "hands on" personnel associated with smaller-sized chemical manufacturing facilities. The hazardous substances in question number almost 700, excluding oil, and are those designated persuant to Section 101 (14) of the Comprehensive Environmental Response, Compensation and Liability Act of 1980, otherwise known as CERCLA or Superfund (Public Law 96-510).

The earlier Clean Water Act (Public Law 92-500) in Section 311, required that the President issue regulations "establishing procedures, methods and equipment...to prevent discharges of oil and hazardous substances from vessels and from onshore and offshore facilities..." Under a 1973 executive order, the U.S. Environmental Protection Agency (EPA) was to promulgate regulations for preventing nontransportation-related spills.

In 1973, EPA issued oil pollution prevention regulations (40 CFR Part 112) that require certain fixed facilities to have on file a spill prevention, control and countermeasures plan (SPCC plan). These regulations have been largely accepted by industry and have proven to be effective in preventing oil spills.

In the case of hazardous substances, no similar federal regulations exist at this time. In its own interest, a large segment of the industry producing, storing, and handling hazardous substances has instituted internal spill prevention plans. However, many of the smaller affected facilities may not have generated spill prevention plans for various reasons. This manual is designed to assist them in preparing such plans.

Source material for this manual was derived from government, industry and commercial publications. Related oil spill prevention literature included: the Guide for Inspectors (11); a report on Prevention Practices at Small Petroleum Facilities (10); state contingency plans, such as the Oil Spill Contingency Plan of the State of California (9); and the extensive Oil Spill Prevention Control and Countermeasure Plan Review (7). Publications dealing with hazardous substances included cost analyses for hazardous substance pollution prevention (5, 6), Best Management Practices (BMP) documents (1, 3), an industrial spill prevention plan (2), and a treatise on Safety and Accident Prevention in Chemical Operations (4).

The manual consists of seven sections and an Appendix: 1. Introduction; 2. Manual of Practice; 3. Hazardous Substances and Their Characteristics; 4. Fixed Facilities; 5. Facility Spill Prevention Practices; 6. Preventive Engineering Practices; and 7. Bibliography. Section 2, the manual proper, deals with the preparation of a Facility Spill Prevention Master Plan. To this end, the facility is evaluated for areas and equipment items which interact with hazardous substances--be they raw materials, products, by-products, or wastes. A Spill Prevention Committee oversees the preparation of the plan. Section 3 contains several listings of hazardous substances in accordance with their physical state upon release under ambient conditions, specific hazards, and behavior on release into water. Section 4 includes a listing of major fixed facility chemical processing equipment elements that would interact with hazardous substances. The Appendix contains descriptions of the equipment components for use in preparing a checklist of equipment items which interact with hazardous substances. Section 5 lists the various spill prevention practices including, organization, risk identification, material compatibility, preventive maintenance, good housekeeping, security and training. Section 6 details preventive engineering concepts such as monitoring and secondary containment of storage vessels, fire protection systems, alarm systems, valving and venting, drainage control, waste treatment, etc. Section 7 lists the sources used in preparing the manual.

2. Manual of Practice

This section presents the procedure for developing a facility spill prevention master plan. The procedure is divided into 10 steps which are expanded in the remainder of the manual. The manager selected to oversee the plan, and constant support of upper level management, are essential for successful implementation of the project. Specialists as required to perform the various tasks outlined below, and the involvement of people from different parts of the facility, will provide a wide base of support for the plan and increase its acceptance by plant personnel who will be its main beneficiaries. The 10 procedural steps are as follows:

1. Form a facility spill prevention organization supervised by a part-time or full-time manager and backed by the plant management.

2. Prepare a prevention policy statement approved by management.

3. Define facility boundaries and prepare flowsheets indicating generation and storage of all substances within these boundaries, as well as inflow and outflow of substances across the boundaries. Hazardous substances present in raw materials, products, byproducts, wastes, fuels, lubes, paints, pesticides, disinfectants, etc., should be identified. Normal and overload conditions should be noted.

4. List the substances that are hazardous by reference to Section 3 and the tables of hazardous substance characteristics contained therein. Note the environmental media into which the substances would be released, their physical behavior on release, and the hazards caused by the release.

5. List all facility areas and equipment items that interact with hazardous substances. Include storage vessels of all kinds (gas, liquid, solid), process vessels and columns, flow systems including valves and controls, receiving and shipping terminals of all kinds (road, rail, water, air), and waste treatment and disposal areas. Use Section 4, Fixed Facilities, to identify those areas and equipment that could interact with hazardous substances.

6. For each area and equipment item of interest, list possible failure modes; amount of hazardous substance involved; hazards caused by possible release of substance (from Step 4 above); and specific effects expected on the rest of the facility and surroundings,

considering equipment and personnel at various times and varying weather conditions.

7. For each area and equipment item, look over Section 5, Facility Spill Prevention Practices (SPPs), and extract the applicable Preventive Engineering Practices.

8. For each area and equipment item, look over Section 6, Preventive Engineering Practices (PEPs), and extract the applicable PEPs.

9. Write a facility spill prevention master plan under direction of the manager by combining information from Steps 3 through 8. The plan should consider drainage to receiving waters and facility terrain, and should include a timetable for carrying out the SPPs and PEPs.

10. The plan should be approved by plant management and implemented under direction of the spill prevention organization which also should be responsible for its periodic review and revision.

3. Hazardous Substances and Their Characteristics

This section contains tables of information on the almost 700 CERCLA-designated hazardous substances. Once all the substances within a plant have been identified, they should be checked against these tables to determine which are hazardous. Only those considered hazardous require further consideration.

Hazardous substances may be gases, liquids, or solids; and, they may be released to the air, water, or ground. Four tables are thus provided: (1) liquids and solids spilled on water, (2) liquids spilled on ground, (3) solids (particulates) released to air or ground, and (4) gases released to air.

Table 1 has five columns: (1) alphabetical listing of substances, primarily chemical compounds; (2) chemical class, sometimes more than one; (3) Chemical Abstract Service (CAS) number, a standard in cases where substances are known by more than one name; (4) hazards in addition to toxicity; and (5) behavior in water - sink/float and soluble/insoluble. The fifth or last column, "Behavior in Water," is not included in Tables 2, 3, and 4.

The hazards listed in the fourth column of each table are defined in 49 CFR 173 (Department of Transportation (DOT) regulations), the Clean Water Act, and by the U.S. Department of Health and Human Services. In the following list, an asterisk designates definitions developed during this work.

- <u>Carcinogen</u> - substance identified as potentially cancer-producing in humans.
- <u>Combustible</u> - liquid or solid having a flash point at or above 100°F and below 300°F; the upper limit was changed from the DOT value of 200°F to 300°F to realistically include more substances as combustible.
- <u>Corrosive</u> - substance causing visible destruction or irreversible alterations in human skin tissue at the site of contact.
- <u>Explosive</u> - any chemical compound, mixture, or device providing substantial instantaneous release of gas and heat.
- <u>Flammable</u> - substance as defined in 49 CFR 173.300 (gas), 49 CFR 173.115 (liquid), and 49 CFR 173.150 (solid).
- <u>Oxidizer</u> - substance that yields oxygen readily to stimulate the combustion of organic matter.

- Poison - substance so classed or labelled in 49 CFR 172.101.

- Polymerizable* - substance undergoing a rapid exothermic polymerization reaction initiated by exposure to heat, light, acids, caustics, or other sources.

- Radioactive material - substance spontaneously emitting ionizing radiation.

- Reactive* - substance that readily undergoes violent change without detonation in the presence of water or moist air, or even dry air or oxygen.

- Toxic pollutant- material which upon exposure, ingestion, inhalation, or assimilation into any organism, causes death, disease, behavioral abnormalities, cancer, genetic mutations, physiological malfunctions, or physical deformations.

A thorough understanding of the various hazards is essential for effective selection of preventive countermeasures. This section provides data in the form of four tables of Hazardous Substance Characteristics for identifying and listing those substances that are hazardous in accordance with Step 4 of the procedures for developing a facility spill prevention master plan.

Table #	Release Scenario	Pages
1	Releases in Water	7 - 46
2	Liquids Released on Land	47 - 60
3	Particulate Solids Released on Land	61 - 83
4	Compressed Gases Released Into Air	84 - 85

TABLE 1. RELEASES IN WATER

Hazardous Substance	Chemical Class	CAS No.	Hazard(s), in Addition to Toxicity	Behavior in Water
Acenaphthene	Aromatics	83-32-9	Combustible	Insoluble Sinker
Acenaphthylene	Aromatics	208-96-8	Combustible	Insoluble Floater
Acetaldehyde	Aldehydes	75-07-0	Flammable Polymerizable	Soluble
Acetic acid	Acidic compounds, organic	64-19-7	Combustible Corrosive	Soluble
Acetic anhydride	Acidic compounds, organic	108-24-7	Combustible Corrosive	Soluble, decomposes
Acetone	Ketones	67-64-1	Flammable	Soluble
Acetone cyanohydrin	Cyanides and nitriles	75-86-5	Combustible w/toxic products Poison	Soluble
Acetonitrile	Cyanides and nitriles	75-05-8	Flammable w/toxic products	Soluble
Acetophenone	Ketones	98-86-2	Combustible	Insoluble Sinker
Acetyl bromide	Aliphatics, halogenated	506-96-7	Flammable w/toxic products Corrosive Reactive	Decomposes (Sinker)
Acetyl chloride	Aliphatics, halogenated	75-36-5	Flammable w/toxic products Corrosive Reactive	Decomposes (Sinker)
2-Acetylaminofluorene	Amines, aryl	53-96-3	Potential carcinogen	Insoluble Sinker
1-Acetyl-2-thiourea	Ureas	591-08-2		Soluble
Acrolein	Aldehydes, Olefins	107-02-8	Flammable Polymerizable Poison	Soluble
Acrylamide	Amides, anilides, and imides	79-06-1	Polymerizable	Soluble

TABLE 1. RELEASES IN WATER

Hazardous Substance	Chemical Class	CAS No.	Hazard(s), in Addition to Toxicity	Behavior in Water
Acrylic acid	Acidic compounds, organic Olefins	79-10-7	Combustible Corrosive Polymerizable	Soluble
Acrylonitrile	Cyanides and nitriles,	107-13-1	Flammable w/toxic products Polymerizable Potential carcinogen Poison	Soluble
Adipic acid	Acidic compounds, organic	124-04-9		Soluble
Aldicarb	Esters	116-06-3		Insoluble Sinker
Aldrin	Aromatics, halogenated	309-00-2	Combustible w/toxic products Potential carcinogen Poison	Insoluble Sinker
Allyl alcohol	Alcohols and glycols, Olefins	107-18-6	Flammable Poison	Soluble
Allyl chloride	Halides, alkyl, Olefins	107-05-1	Flammable w/toxic products	Insoluble Floater
Aluminum phosphide	Phosphorous and compounds	20859-73-8	Flammable w/toxic products Reactive	Liberates poisonous phosphine on contact
Aluminum sulfate	Sulfates	10043-01-3		Soluble
5-(Aminomethyl)-3-isoxazole (Note 1)	Amines, alkyl	2763-96-4		Soluble
Amitrole (Note 1)	Azo compounds	61-82-5	Potential carcinogen	Soluble
Ammonium acetate	Organic ammonium compounds	631-61-8		Soluble
Ammonium benzoate	Organic ammonium compounds	1863-63-4	Combustible w/toxic products	Soluble
Ammonium bicarbonate	Organic ammonium compounds	1066-33-7		Soluble

TABLE 1. RELEASES IN WATER

Hazardous Substance	Chemical Class	CAS No.	Hazard(s), in Addition to Toxicity	Behavior in Water
Ammonium bichromate	Chromates	7789-09-5	Corrosive Oxidizer Flammable	Soluble
Ammonium bifluoride	Halides, inorganic	1341-49-7	Corrosive	Soluble
Ammonium bisulfite	Sulfites	10192-30-0		Soluble
Ammonium carbamate	Esters	1111-78-0		Soluble
Ammonium carbonate	Organic ammonium compounds	10361-29-2		Soluble
Ammonium chloride	Halides, inorganic	12125-02-9		Soluble
Ammonium chromate	Chromates	7788-98-9		Soluble
Ammonium citrate, dibasic	Organic ammonium compounds	3012-65-5		Soluble
Ammonium fluoborate	Organic ammonium compounds	13826-83-0	Corrosive	Soluble
Ammonium fluoride	Halides, inorganic	12125-01-8	Corrosive	Soluble
Ammonium hydroxide	Basic compounds	1336-21-6	Corrosive	Soluble
Ammonium oxalate	Organic ammonium compounds	5972-73-6 14258-49-2 6009-70-7		Soluble
Ammonium picrate	Nitro compounds	131-74-8	Flammable w/toxic products Explosive	Soluble
Ammonium silicofluoride	Halides, inorganic	16919-19-0	Corrosive	Soluble
Ammonium sulfamate	Sulfones, sulfoxides, and sulfonates	7773-06-0		Soluble
Ammonium sulfide	Sulfides and mercaptans	12135-76-1	Flammable w/toxic products	Soluble

TABLE 1. RELEASES IN WATER

Hazardous Substance	Chemical Class	CAS No.	Hazard(s), in Addition to Toxicity	Behavior in Water
Ammonium sulfite	Sulfites	10196-04-0	Combustible w/toxic products	Soluble
Ammonium tartrate	Organic ammonium compounds	3164-29-2 14307-43-8		Soluble
Ammonium thiocyanate	Cyanates	1762-95-4	Combustible w/toxic products	Soluble
Ammonium thiosulfate	Sulfates	7783-18-8		Soluble
Ammonium vanadate	Heavy metals	7803-55-6		Insoluble Sinker
Amyl acetate	Esters	628-63-7	Flammable	Insoluble Floater
Aniline	Amines, aryl	62-53-3	Combustible w/toxic products Poison	Soluble
Anthracene	Aromatics	120-12-7	Combustible	Insoluble Sinker
Antimony	Heavy metals	7440-36-0	Combustible w/toxic products	Insoluble Sinker
Antimony pentachloride	Halides, inorganic Heavy metals	7647-18-9	Corrosive Reactive	Reacts vigorously, liberating HCl
Antimony potassium tartrate	Organometallics Heavy metals	28300-74-5		Insoluble Sinker
Antimony tribromide	Halides, inorganic Heavy metals	7789-61-9	Corrosive Reactive	Insoluble Sinker
Antimony trichloride	Halides, inorganic Heavy metals	10025-91-9	Corrosive Reactive	Reacts vigorously, liberating HCl
Antimony trifluoride	Halides, inorganic Heavy metals	7783-56-4	Corrosive Reactive	Soluble
Antimony trioxide	Oxides Heavy metals	1309-64-4		Insoluble Sinker

TABLE 1. RELEASES IN WATER

Hazardous Substance	Chemical Class	CAS No.	Hazard(s), in Addition to Toxicity	Behavior in Water
Arsenic	Heavy metals	7440-38-2	Combustible w/toxic products Potential carcinogen Poison	Insoluble Sinker
Arsenic acid	Acidic compounds, inorganic Heavy metals	7778-39-4 1327-52-2	Corrosive Poison	Soluble
Arsenic disulfide	Sulfides and mercaptans Heavy metals	1303-32-8	Combustible w/toxic products Poison	Insoluble Sinker
Arsenic pentoxide	Oxides Heavy metals	1303-28-2	Corrosive Poison	Soluble
Arsenic trichloride	Halides, inorganic Heavy metals	7784-34-1	Corrosive Reactive Poison	Decomposes (Sinker)
Arsenic trioxide	Oxides Heavy metals	1327-53-3	Corrosive Poison	Soluble
Arsenic trisulfide	Sulfides and mercaptans Heavy metals	1303-33-9	Combustible w/toxic products Poison	Insoluble Sinker
Asbestos	(See asbestos)	1332-21-4	Potential carcinogen	Insoluble Sinker
Auramine	Amines, aryl	492-80-8	Potential carcinogen	Insoluble Sinker
Azaserine (Note 1)	Azo compounds	115-02-6	Potential carcinogen	Soluble
Barium cyanide	Cyanides and nitriles	542-62-1	Poison	Soluble
3,4-Benzacridine	Aromatics	225-51-4		Insoluble Sinker
Benzal chloride	Aromatics, halogenated	98-87-3		Insoluble Sinker
1,2-Benzanthracene (Note 1)	Aromatics	56-55-3	Potential carcinogen	Insoluble Sinker

TABLE 1. RELEASES IN WATER

Hazardous Substance	Chemical Class	CAS No.	Hazard(s), in Addition to Toxicity	Behavior in Water
Benzene	Aromatics	71-43-2	Flammable Potential carcinogen	Insoluble Floater
Benzenesulfonyl chloride	Acidic compounds, organic	98-09-9	Combustible w/toxic products	Insoluble Sinker
Benzidine	Amines, aryl	92-87-5	Combustible w/toxic products Potential carcinogen Poison	Insoluble Sinker
Benzo[k]fluoranthene	Aromatics	207-08-9	Potential carcinogen	Insoluble Sinker
Benzo[b]fluoranthene	Aromatics	205-99-2	Potential carcinogen	Insoluble Sinker
Benzoic acid	Acidic compounds, organic	65-85-0	Combustible	Insoluble Sinker
Benzonitrile	Cyanides and nitriles	100-47-0	Combustible w/toxic products	Soluble
Benzo[ghi]perylene	Aromatics	191-24-2		Insoluble Sinker
Benzo[a]pyrene	Aromatics	50-32-8	Potential carcinogen	Insoluble Sinker
p-Benzoquinone	Ketones	106-51-4	Combustible	Soluble
Benzotrichloride	Aromatics,halogenated	98-07-7	Combustible w/toxic products Corrosive	Decomposes (Sinker)
Benzoyl chloride	Aromatics, halogenated	98-88-4	Combustible w/toxic products Corrosive	Decomposes (Sinker)
Benzyl chloride	Aromatics, halogenated	100-44-7	Combustible w/toxic products Corrosive Reactive	Insoluble Sinker
Beryllium	Heavy metals	7440-41-7	Flammable w/toxic products Potential carcinogen	Insoluble Sinker

TABLE 1. RELEASES IN WATER

Hazardous Substance	Chemical Class	CAS No.	Hazard(s), in Addition to Toxicity	Behavior in Water
Beryllium chloride	Halides, inorganic Heavy metals	7787-47-5	Poison	Soluble
Beryllium fluoride	Halides, inorganic Heavy metals	7787-49-7	Poison	Soluble
Beryllium nitrate	Nitrates and nitrites Heavy metals	7787-55-5 13597-99-4	Oxidizer	Soluble
alpha-Benzenehexa-chloride	Aliphatics, halogenated	319-84-6	Potential carcinogen	Insoluble Sinker
beta-Benzenehexa-chloride	Aliphatics, halogenated	319-85-7	Potential carcinogen	Insoluble Sinker
delta- Benzenehexa-chloride	Aliphatics, halogenated	319-86-8	Potential carcinogen	Insoluble Sinker
2,2'-Bioxirane	Epoxides	1464-53-5	Potential carcinogen	Decomposes (Sinker)
Bis(2-chloroethoxy)-methane	Aliphatics, halogenated	111-91-1		Soluble
Bis(2-chloroethyl) ether	Ethers	111-44-4	Combustible w/toxic products Poison	Soluble
Bis(2-chloroisopropyl)-ether	Ethers Aliphatics, halogenated	108-60-1	Combustible w/toxic products	Insoluble Sinker
Bis(chloromethyl) ether	Ethers	542-88-1	Combustible w/toxic products Potential carcinogen	Soluble
Bis(2-ethylhexyl) phthalate	Esters	117-81-7		Insoluble Floater
Bromoacetone (Note 2)	Ketones	598-31-2	Poison	Soluble
Bromoform	Halides, alkyl	75-25-2		Insoluble Sinker
4-Bromophenyl phenyl ether	Ethers Aromatics, halogenated	101-55-3	Combustible w/toxic products	Insoluble Sinker

TABLE 1. RELEASES IN WATER

Hazardous Substance	Chemical Class	CAS No.	Hazard(s), in Addition to Toxicity	Behavior in Water
Brucine	(See strychnine and salts)	357-57-3	Combustible w/toxic products Poison	Insoluble Sinker
1-Butanol	Alcohols and glycols	71-36-3	Flammable	Soluble
2-Butanone peroxide (Note 2)	Peroxides	1338-23-4	Explosive Oxidizer Combustible	Soluble
Butyl acetate	Esters	123-86-4	Flammable	Soluble
Butyl benzyl phthalate	Esters	85-68-7		Insoluble Sinker
Butylamine	Amines, alkyl	109-73-9	Flammable w/toxic products	Soluble
Butyric acid	Acidic compounds, organic	107-92-6	Combustible	Soluble
Cacodylic acid	Organometallics Heavy metals	75-60-5	Poison	Soluble
Cadmium	Heavy metals	7440-43-9	Flammable w/toxic products Potential carcinogen	Insoluble Sinker
Cadmium acetate	Organometallics Heavy metals	543-90-8	Poison	Soluble
Cadmium bromide	Halides, inorganic Heavy metals	7789-42-6	Poison	Soluble
Cadmium chloride	Halides, inorganic Heavy metals	10108-64-2	Potential carcinogen	Soluble
Calcium arsenate	Heavy metals	7778-44-1	Poison	Insoluble Sinker
Calcium arsenite	Heavy metals	52740-16-6	Poison	Insoluble Sinker
Calcium carbide	Organometallics	75-20-7	Flammable Reactive	Inflames on contact

TABLE 1. RELEASES IN WATER

Hazardous Substance	Chemical Class	CAS No.	Hazard(s), in Addition to Toxicity	Behavior in Water
Calcium chromate	Chromates	13765-19-0	Potential carcinogen	Insoluble Sinker
Calcium cyanide	Cyanides and nitriles	592-01-8	Reactive Poison	Forms toxic cyanide
Calcium dodecylbenzene sulfonate	Sulfones, sulfoxides and sulfonates	26264-06-2		Insoluble Sinker
Calcium hypochlorite	Halides, inorganic	7778-54-3	Oxidizer	Insoluble Sinker
Captan	Acidic compounds, organic Amides, anilides, and imides	133-06-2	Combustible w/toxic products	Insoluble Sinker
Carbaryl	Esters	63-25-2	Combustible w/toxic products	Insoluble Sinker
Carbofuran	Esters	1563-66-2	Combustible w/toxic products Poison	Insoluble Sinker
Carbon disulfide	Sulfides and mercaptans	75-15-0	Flammable w/toxic products	Insoluble Sinker
Carbon tetrachloride	Halides, alkyl	56-23-5	Potential carcinogen	Insoluble Sinker
Chloral	Aldehydes Aliphatics,halogenated	75-87-6	Combustible w/toxic products Corrosive Potential carcinogen	Soluble
Chlorambucil	Aromatics, halogenated Amines, aryl	305-03-3	Potential carcinogen	Insoluble Sinker
p-Chloro-m-cresol	Phenols and cresols Aromatics, halogenated	59-50-7	Combustible w/toxic products	Insoluble Sinker
Chlordane	Aliphatics, halogenated	57-74-9	Combustible w/toxic products Potential carcinogen	Insoluble Sinker
Chlornaphazine (Note 1)	Amines, aryl	494-03-1	Potential carcinogen	Insoluble Sinker

TABLE 1. RELEASES IN WATER

Hazardous Substance	Chemical Class	CAS No.	Hazard(s), in Addition to Toxicity	Behavior in Water
Chloroacetaldehyde	Aldehydes Aliphatics, halogenated	107-20-0	Combustible w/toxic products Polymerizable	Soluble
p-Chloroaniline	Aromatics, halogenated Amines, aryl	106-47-8		Soluble
Chlorobenzene	Aromatics, halogenated	108-90-7	Flammable w/toxic products	Insoluble Sinker
Chlorodibromomethane	Aliphatics, halogenated	124-48-1		Insoluble Sinker
Chloroethane	Aliphatics, halogenated	75-00-3	Flammable w/toxic products	Insoluble Floater
2-Chloroethyl vinyl ether	Ethers Aliphatics, halogenated	110-75-8	Flammable w/toxic products	Soluble
Chloroform	Halides, alkyl	67-66-3	Potential carcinogen	Insoluble Sinker
Chloromethyl methyl ether	Ethers Aliphatics, halogenated	107-30-2	Flammable w/toxic products Potential carcinogen Poison	Insoluble Sinker
2-Chloronaphthalene	Aromatics, halogenated	91-58-7	Combustible w/toxic products	Insoluble Sinker
2-Chlorophenol	Aromatics, halogenated Phenols and cresols	95-57-8	Combustible w/toxic products	Soluble
4-Chlorophenyl phenyl ether	Ethers, Aromatics, halogenated	7005-72-3	Combustible w/toxic products	Insoluble Sinker
3-Chloropropionitrile	Cyanides and nitriles	542-76-7	Combustible w/toxic products	Soluble
Chlorosulfonic acid	Acidic compounds, organic	7790-94-5	Corrosive Reactive	Dangerously reacts giving off HCl and H_2SO_4
4-Chloro-o-toluidine, hydrochloride	Aromatics, halogenated Amines, aryl	3165-93-3	Poison	Soluble
Chlorpyrifos	Organophosphates Aromatics, halogenated	2921-88-2	Combustible w/toxic products	Insoluble Sinker

TABLE 1. RELEASES IN WATER

Hazardous Substance	Chemical Class	CAS No.	Hazard(s), in Addition to Toxicity	Behavior in Water
Chromic acetate	Organometallics	1066-30-4		Soluble
Chromic acid	Acidic compounds, inorganic	11115-74-5	Corrosive Oxidizer Potential carcinogen	Soluble
Chromic sulfate	Sulfates	10101-53-8		Soluble
Chromium	Heavy metals	7440-47-3	Flammable w/toxic products	Insoluble Sinker
Chromous chloride	Halides, inorganic	10049-05-5	Reactive	Soluble
Chrysene	Aromatics	218-01-9	Combustible	Insoluble Sinker
Cobaltous bromide	Halides, inorganic Heavy metals	7789-43-7		Soluble
Cobaltous formate	Organometallics Heavy metals	544-18-3		Soluble
Cobaltous sulfamate	Heavy metals	14017-41-5		Insoluble Sinker
Copper	Heavy metals	7440-50-8		Insoluble Sinker
Copper (cuprous) cyanide	Cyanides and nitriles Heavy metals	544-92-3	Poison	Insoluble Sinker
Coumphos	Organophosphates Aromatics, halogenated	56-72-4	Combustible w/toxic products Poison	Insoluble Sinker
Creosote	Phenols and Cresols	8001-58-9	Combustible Potential carcinogen	Insoluble Sinker
Cresol	Phenols and cresols	1319-77-3	Combustible	Soluble
Crotonaldehyde	Aldehydes, Olefins	4170-30-3 123-73-9	Flammable	Soluble

TABLE 1. RELEASES IN WATER

Hazardous Substance	Chemical Class	CAS No.	Hazard(s), in Addition to Toxicity	Behavior in Water
Cumene	Aromatics	98-82-8	Combustible	Insoluble Floater
Cupric acetate	Organometallics Heavy metals	142-71-2		Soluble
Cupric acetoarsenite	Organometallics Heavy metals	12002-03-8	Poison	Insoluble Sinker
Cupric chloride	Halides, inorganic Heavy metals	7447-39-4		Soluble
Cupric nitrate	Nitrates and nitrites Heavy metals	3251-23-8		Soluble
Cupric oxalate	Organometallics Heavy metals	814-91-5		Insoluble Sinker
Cupric sulfate	Sulfates Heavy metals	7758-98-7		Soluble
Cupric sulfate ammoniated	Sulfates Heavy metals	10380-29-7		Soluble
Cupric tartrate	Organometallics Heavy metals	815-82-7		Insoluble Sinker
Cyanides (soluble salts and complexes)	Cyanides and nitriles	57-12-5	Poison	Soluble
Cyanogen bromide	Cyanides and nitriles	506-68-3	Poison	Soluble
Cyanogen chloride	Cyanides and nitriles	506-77-4	Poison	Reacts to give toxic cyanide ion
Cyclohexane	Aliphatics	110-82-7	Flammable	Insoluble Floater
Cyclohexanone	Ketones	108-94-1	Combustible	Soluble
Cyclophosphamide (Note 2)	Organophosphates, Amides, anilides, and imides	50-18-0	Potential carcinogen	Soluble
2,4-D acid	Acids, organic Aromatics, halogenated	94-75-7	Combustible w/toxic products	Insoluble Sinker

TABLE 1. RELEASES IN WATER

Hazardous Substance	Chemical Class	CAS No.	Hazard(s), in Addition to Toxicity	Behavior in Water
2,4-D esters	Esters	94-11-1	Combustible w/toxic products	Insoluble Sinker
Daunomycin (Note 1)	Aromatics Ketones	20830-81-3	Potential carcinogen	Soluble
DDD	Aromatics, halogenated	72-54-8	Combustible w/toxic products Potential carcinogen	Insoluble Sinker
DDE	Aromatics, halogenated	72-55-9	Combustible w/toxic products Potential carcinogen	Insoluble Sinker
DDT	Aromatics, halogenated	50-29-3	Combustible w/toxic products Potential carcinogen	Insoluble Sinker
Diallate	Esters	2303-16-4		Insoluble Sinker
Diazinon	Organophosphates	333-41-5	Combustible w/toxic products	Insoluble Sinker
Dibenzo[a,h]anthracene	Aromatics	53-70-3	Potential carcinogen	Insoluble Sinker
Dibenz[a,i]pyrene	Aromatics	189-55-9	Potential carcinogen	Insoluble Sinker
1,2-Dibromo-3-chloropropane (Note 2)	Aliphatics, halogenated	96-12-8	Combustible w/toxic products Potential carcinogen	Insoluble Sinker
Di-n-butylphthalate	Esters	84-74-2		Insoluble Sinker
Dicamba	Acidic compounds, organic Aromatics, halogenated	1918-00-9	Combustible w/toxic products	Insoluble Sinker
Dichlobenil	Cyanides and nitriles Aromatics, halogenated	1194-65-6	Combustible w/toxic products	Insoluble Sinker
Dichlone	Aromatics, halogenated	117-80-6	Combustible w/toxic products	Insoluble Sinker

TABLE 1. RELEASES IN WATER

Hazardous Substance	Chemical Class	CAS No.	Hazard(s), in Addition to Toxicity	Behavior in Water
Dichlorobenzene (all isomers)	Aromatics, halogenated	25321-22-6	Combustible w/toxic products	Insoluble Sinker
o-Dichlorobenzene	Aromatics, halogenated	95-50-1	Combustible w/toxic products	Insoluble Sinker
m-Dichlorobenzene	Aromatics, halogenated	541-73-1	Combustible w/toxic products	Insoluble Sinker
p-Dichlorobenzene	Aromatics, halogenated	106-46-7	Combustible w/toxic products	Insoluble Sinker
3,3'-Dichlorobenzidine	Aromatics, halogenated Amines,aryl	91-94-1	Combustible w/toxic products Potential carcinogen	Insoluble Sinker
Dichlorobromomethane	Aliphatics, halogenated	75-27-4	Combustible w/toxic products	Insoluble Sinker
1,4-Dichloro-2-butene	Aliphatics, halogenated	764-41-0	Flammable w/toxic products	Insoluble Sinker
1,1-Dichloroethane	Aliphatics, halogenated	75-34-3	Flammable w/toxic products	Insoluble Sinker
1,2-trans-Dichloro-ethylene	Aliphatics, halogenated	156-60-5	Flammable w/toxic products	Insoluble Sinker
Dichloropropane (all isomers)	Aliphatics, halogenated	26638-19-7	Flammable w/toxic products	Insoluble Sinker
1,3-Dichloropropene	Aliphatics, halogenated	542-75-6	Flammable w/toxic products	Insoluble Sinker
Dichloropropene (all isomers)	Aliphatics, halogenated	26952-23-8	Flammable w/toxic products	Insoluble Sinker
Dichloropropene-Dichloropropane mixture	Aliphatics, halogenated	8003-19-8	Flammable w/toxic products	Insoluble Sinker
2,2-Dichloropropionic acid	Aliphatics, halogenated	75-99-0	Corrosive	Soluble
Dichlorvos	Organophosphates	62-73-7	Poison	Insoluble Sinker
2,4-Dichlorophenol	Phenol and cresols Aromatics, halogenated	120-83-2	Combustible w/toxic products	Insoluble Sinker

TABLE 1. RELEASES IN WATER

Hazardous Substance	Chemical Class	CAS No.	Hazard(s), in Addition to Toxicity	Behavior in Water
2,6-Dichlorophenol	Phenols and cresols Aromatics, halogenated	87-65-0	Combustible w/toxic products	Insoluble Sinker
Dieldrin	Epoxides Aromatics, halogenated	60-57-1	Potential carcinogen	Insoluble Sinker
Diethylamine	Amines, alkyl	109-89-7	Flammable w/toxic products Corrosive	Soluble
Diethylarsine (Note 2)	Organometallics Heavy metals	692-42-2	Flammable w/toxic products	Insoluble Sinker
N,N'-Diethylhydrazine	Hydrazines and hydrazides	1615-80-1	Combustible w/toxic products Potential carcinogen	Soluble
Diethylstilbesterol	Aromatics Ketones	56-53-1	Potential carcinogen	Insoluble Sinker
O,O-Diethyl-S-methyl dithiophosphate (Note 2	Organophosphates	3288-58-2		Insoluble Sinker
Diethyl-p-nitrophenyl phosphate	Organophosphates Nitro compounds	311-45-5		Soluble
Diethyl phthalate	Esters	84-66-2		Insoluble Sinker
O,O-Diethyl O-pyrazinyl phosphorothioate	Organophosphates	297-97-2		Soluble
Dihydrosafrole (Note 2)	Aromatics Ethers	94-58-6	Potential carcinogen	Insoluble Sinker
Diisopropyl fluorophos-phate	Organophosphates	55-91-4	Corrosive	Soluble
Dimethoate	Organophosphates	60-51-5	Flammable w/toxic products	Soluble
3,3'-Dimethoxybenzidene	Amines, aryl	119-90-4	Potential carcinogen	Insoluble Sinker
alpha,alpha-Dimethyl-benzylhydroperoxide	Peroxides Aromatics	80-15-9	Explosive Combustible	Insoluble Sinker
Dimethyl carbamoyl chloride (Note 2)	Halides, alkyl	79-44-7	Combustible w/toxic products Reactive Potential carcinogen	Soluble

TABLE 1. RELEASES IN WATER

Hazardous Substance	Chemical Class	CAS No.	Hazard(s), in Addition to Toxicity	Behavior in Water
alpha,alpha-Dimethylphenethylamine (Note 1)	Amines, aryl	122-09-8		Soluble
Dimethyl phthalate	Esters	131-11-3	Combustible	Insoluble Sinker
Dimethyl sulfate	Sulfates	77-78-1	Combustible w/toxic products Corrosive Potential carcinogen	Soluble
Dimethylaminoazobenzene	Azo compounds Amines, aryl	60-11-7	Potential carcinogen	Insoluble Sinker
7,12-Dimethylbenz[a]anthracene	Aromatics	57-97-6		Insoluble Sinker
3,3'-Dimethylbenzidine	Amines, aryl	119-93-7	Potential carcinogen	Insoluble Sinker
1,1-Dimethylhydrazine	Hydrazines and hydrazides	57-14-7	Flammable w/toxic products Corrosive Potential carcinogen Poison	Soluble
1,2-Dimethylhydrazine	Hydrazines and hydrazides	540-73-8	Flammable w/toxic products Corrosive Potential carcinogen	Soluble
2,4-Dimethylphenol	Phenols and cresols	105-67-9	Combustible	Insoluble Floater
Dinitrobenzene (mixed)	Nitro compounds Aromatics	25154-54-5	Poison	Insoluble Sinker
4,6-Dinitro-o-cresol	Nitro compounds Phenols and cresols	534-52-1	Combustible w/toxic products	Insoluble Sinker
4,6-Dinitro-o-cyclohexyl phenol	Nitro compounds Phenols and cresols	131-89-5	Combustible w/toxic products	Insoluble Sinker
Dinitrophenol	Nitro compounds Phenols and cresols	25550-58-7	Combustible w/toxic products Poison	Insoluble Sinker

TABLE 1. RELEASES IN WATER

Hazardous Substance	Chemical Class	CAS No.	Hazard(s), in Addition to Toxicity	Behavior in Water
2,4-Dinitrophenol	Nitro compounds Phenols and cresols	51-28-5	Combustible w/toxic products Poison	Insoluble Sinker
2,6-Dinitrotoluene	Nitro compounds Aromatics	606-20-2		Insoluble Sinker
2,4-Dinitrotoluene	Nitro compounds Aromatics	121-14-2		Insoluble Sinker
Dinitrotoluene	Nitro compounds Aromatics	25321-14-6		Insoluble Sinker
Dinoseb	Nitro compounds Phenols and cresols	88-85-7	Flammable w/toxic products	Insoluble Sinker
Di-n-octylphthalate	Esters	117-84-0		Insoluble Floater
1,4-Dioxane	Ethers	123-91-1	Flammable Potential carcinogen	Soluble
2-Diphenylhydrazine	Hydrazines and hydrazides	122-66-7	Combustible w/toxic products Potential carcinogen	Insoluble Sinker
Dipropylamine	Amines, alkyl	142-84-7	Flammable w/toxic products	Soluble
Diquat	Aromatics, halogenated	85-00-7 2764-72-9	Combustible w/toxic products	Soluble
Disulfoton	Organophosphates	298-04-4	Combustible w/toxic products Poison	Insoluble Sinker
2,4-Dithiobiuret	Amides, anilides, and imides	541-53-7		Insoluble Sinker
Diuron	Ureas Aromatics, halogenated	330-54-1	Combustible w/toxic products	Insoluble Sinker

TABLE 1. RELEASES IN WATER

Hazardous Substance	Chemical Class	CAS No.	Hazard(s), in Addition to Toxicity	Behavior in Water
Dodecylbenzenesulfonic acid	Acidic compounds, organic Aromatics	27176-87-0	Combustible w/toxic products	Soluble
Endosulfan	Aromatics, halogenated Sulfones, sulfoxides and sulfonates	115-29-7	Combustible w/toxic products Poison	Insoluble Sinker
alpha-Endosulfan	Aromatics, halogenated Sulfones, sulfoxides and sulfonates	959-98-8	Poison	Insoluble Sinker
beta-Endosulfan	Aromatics, halogenated Sulfones, sulfoxides and sulfonates	33213-65-9	Poison	Insoluble Sinker
Endosulfan sulfate	Aromatics, halogenated Sulfones, sulfoxides and sulfonates	1031-07-8	Combustible w/toxic products	Insoluble Sinker
Endothall (Note 1)	Acidic compounds, organic	145-73-3		Soluble
Endrin	Epoxides Aromatics, halogenated	72-20-8	Poison	Insoluble Sinker
Endrin aldehyde (Note 2)	Aldehydes	930-55-2		Insoluble Sinker
Epichlorohydrin	Ethers Aliphatics, halogenated	106-89-8	Flammable w/toxic products	Insoluble Sinker
Epinephrin	Amines, aryl	51-43-4		Insoluble Sinker
Ethion	Organophosphates	563-12-2	Poison	Insoluble Sinker
Ethyl acetate	Esters	141-78-6	Flammable	Soluble
Ethyl acrylate	Esters, Olefins	140-88-5	Flammable Polymerizable	Soluble
Ethyl carbamate	Esters	51-79-6	Potential carcinogen	Soluble
Ethyl cyanide	Cyanides and nitriles	107-12-0	Flammable w/toxic products	Soluble

TABLE 1. RELEASES IN WATER

Hazardous Substance	Chemical Class	CAS No.	Hazard(s), in Addition to Toxicity	Behavior in Water
Ethyl-4,4'-dichlorobenzilate (Note 1)	Esters Aliphatics, halogenated	510-15-6		Insoluble Sinker
Ethyl ether	Ethers	60-29-7	Explosive, upon standing Flammable	Soluble
Ethyl methacrylate	Esters, Olephins	97-63-2	Flammable Polymerizable	Insoluble Floater
Ethyl methanesulfonate (Note 2)	Esters	62-50-0		Soluble
Ethylbenzene	Aromatics	100-41-4	Flammable	Insoluble Floater
Ethylene dibromide	Aliphatics, halogenated	106-93-4	Potential carcinogen	Insoluble Sinker
Ethylene dichloride	Aliphatics, halogenated	107-06-2	Flammable w/toxic products Potential carcinogen	Insoluble Sinker
Ethylene oxide	Oxides, alkylene	75-21-8	Flammable Corrosive	Soluble
Ethylenebis(dithiocarbamic acid) (Note 1)	Acidic compounds, organic	111-54-6		Soluble
Ethylenediamine	Amines, alkyl	107-15-3	Flammable w/toxic products Corrosive	Soluble
Ethylenediamine tetraacetic acid	Amines, alkyl Acidic compounds, organic	60-00-4	Combustible w/toxic products	Insoluble Sinker
Ethylenethiourea (Note 1)	Ureas	96-45-7	Potential carcinogen	Soluble
Ethylenimine	Amines, alkyl	151-56-4	Flammable w/toxic products Poison	Soluble
Famphur (Note 2)	Organophosphates Amides, anilides, and imides	52-85-7		Insoluble Sinker

TABLE 1. RELEASES IN WATER

Hazardous Substance	Chemical Class	CAS No.	Hazard(s), in Addition to Toxicity	Behavior in Water
Ferric ammonium citrate	Organometallics	1185-57-5		Soluble
Ferric ammonium oxalate	Organometallics	55488-87-4 2944-67-4		Soluble
Ferric chloride	Halides, inorganic	7705-08-0		Soluble
Ferric fluoride	Halides, inorganic	7783-50-8		Insoluble Sinker
Ferric nitrate	Nitrates and nitrites	10421-48-4	Oxidizer	Soluble
Ferric sulfate	Sulfates	10028-22-5		Soluble
Ferrous ammonium sulfate	Sulfates	10045-89-3		Soluble
Ferrous chloride	Halides, inorganic	7758-94-3		Soluble
Ferrous sulfate	Sulfates	7782-63-0 7720-78-7		Soluble
Fluoranthene	Aromatics	206-44-0		Insoluble Sinker
Fluorene	Aromatics	86-73-7		Insoluble Sinker
Fluoroacetamide (Note 1)	Amides, anilides, and imides	640-19-7		Soluble
Formic acid	Acidic compounds, organic	64-18-6	Combustible	Soluble
Fumaric acid	Acidic compounds, organic	110-17-8	Combustible	Insoluble Sinker
Furan	Ethers, Aromatics	110-00-9	Flammable	Soluble
Furfural	Aldehydes Olefins	98-01-1	Combustible	Soluble

TABLE 1 RELEASES IN WATER

Hazardous Substance	Chemical Class	CAS No.	Hazard(s), in Addition to Toxicity	Behavior in Water
Glycidaldehyde (Note 2)	Aldehydes	765-34-4	Flammable Potential carcinogen	Soluble
Guthion	Organophosphates Aromatics	86-50-0	Poison	Insoluble Sinker
Heptachlor	Aliphatics, halogenated	76-44-8	Potential carcinogen	Insoluble Sinker
Heptachlor epoxide	Aliphatics, halogenated Epoxides	1024-57-3	Potential carcinogen	Insoluble Sinker
Hexachlorobenzene	Aromatics, halogenated	118-74-1	Potential carcinogen	Insoluble Sinker
Hexachlorobutadiene	Aliphatics, halogenated Olefins	87-68-3		Insoluble Sinker
Hexachlorocyclo-pentadiene	Aliphatics, halogenated Olefins	77-47-4		Insoluble Sinker
Hexachloroethane	Aliphatics, halogenated	67-72-1		Insoluble Sinker
Hexachlorohexahydro-endo endo-dimethanonaphtha-lene	Aromatics, halogenated	465-73-6	Poison	Insoluble Sinker
Hexachlorophene	Aromatics, halogenated	70-30-4		Insoluble Sinker
Hexachloropropene (Note 2)	Aliphatics, halogenated Olefins	1888-71-7		Insoluble Sinker
Hexaethyl tetraphosphate	Organophosphates	757-58-4	Poison	Soluble
Hydrazine	Hydrazines and hydrazides	302-01-2	Flammable w/toxic products Corrosive Potential carcinogen Poison	Soluble
Hydrochloric acid	Acidic compounds, inorganic	7647-01-0	Corrosive Reactive	Soluble

TABLE 1. RELEASES IN WATER

Hazardous Substance	Chemical Class	CAS No.	Hazard(s), in Addition to Toxicity	Behavior in Water
Hydrocyanic acid	Cyanides and nitriles Acidic compounds, inorganic	74-90-8	Flammable w/toxic products Poison	Soluble
Hydrofluoric acid	Acidic compounds, inorganic	7664-39-3	Corrosive Reactive	Soluble
Indeno(1,2,3-cd)pyrene	Aromatics	193-39-5	Potential carcinogen	Insoluble Sinker
Iron dextran	Organometallics	9004-66-4	Potential carcinogen	Soluble
Isobutyl alcohol	Alcohols and glycols	78-83-1	Flammable	Soluble
Isophorone	Ketones	78-59-1	Combustible	Soluble
Isoprene	Olephins	78-79-5	Flammable Polymerizable	Insoluble Floater
Isopropanolamine dodec-tyl benzenesulfonate	Sulfones, sulfoxides, and sulfonates	42504-46-1	Combustible w/toxic products	Insoluble Sinker
Isosafrole (Note 2)	Aromatics Ethers	120-58-1	Potential carcinogen	Insoluble Sinker
Kelthane	Aromatics, halogenated	115-32-2	Combustible w/toxic products	Insoluble Sinker
Kepone	Aliphatics, halogenated Ketones	143-50-0	Combusible w/toxic products Potential carcinogen	Insoluble Sinker
Lasocarpine (Note 2)	Acidic compounds, organic	303-34-4	Potential carcinogen	Insoluble Sinker
Lead	Heavy metals	7439-92-1		Insoluble Sinker
Lead acetate	Organometallics Heavy metals	301-04-2	Potential carcinogen	Soluble
Lead arsenate	Heavy metals	3687-31-8 7784-40-9 7645-25-2 10102-48-4	Poison	Insoluble Sinker

TABLE 1. RELEASES IN WATER

Hazardous Substance	Chemical Class	CAS No.	Hazard(s), in Addition to Toxicity	Behavior in Water
Lead chloride	Halides, inorganic Heavy metals	7758-95-4		Insoluble Sinker
Lead fluoborate	Halides, inorganic Heavy metals	13814-96-5		Soluble
Lead fluoride	Halides, inorganic Heavy metals	7783-46-2		Insoluble Sinker
Lead iodide	Halides, inorganic Heavy metals	10101-63-0		Insoluble Sinker
Lead nitrate	Nitrates and nitrites Heavy metals	10099-74-8	Oxidizer	Soluble
Lead phosphate	Phosphates and phosphonates Heavy metals	7446-27-7	Potential carcinogen	Insoluble Sinker
Lead stearate	Organometallics Heavy metals	7428-48-0 1072-35-1 56189-09-4		Insoluble Sinker
Lead subacetate	Organometallics Heavy metals	1335-32-6	Potential carcinogen	Soluble
Lead sulfate	Sulfates Heavy metals	15739-80-7 7446-14-2		Insoluble Sinker
Lead sulfide	Sulfides and mercaptans Heavy metals	1314-87-0		Forms toxic hydrogen sulfide
Lead thiocyanate	Cyanates Heavy metals	592-87-0		Insoluble Sinker
Lindane	Aliphatics, halogenated	58-89-9	Potential carcinogen	Insoluble Sinker
Lithium chromate	Chromates	14307-35-8		Soluble
Malathion	Organophosphates	121-75-5		Insoluble Sinker
Maleic acid	Acids, organic	110-16-7		Soluble
Maleic anhydride	Acids, organic	108-31-6	Combustible	Soluble

TABLE 1. RELEASES IN WATER

Hazardous Substance	Chemical Class	CAS No.	Hazard(s), in Addition to Toxicity	Behavior in Water
Maleic hydrazide	Hydrazines and hydrazides	123-33-1	Combustible w/toxic products	Insoluble Sinker
Malononitrile	Cyanides and nitriles	109-77-3	Combustible w/toxic products	Soluble
Melphalan	Aromatics, halogenated Amines, aryl	148-82-3	Potential carcinogen	Insoluble Sinker
Mercaptodimethur	Sulfides and mercaptans	2032-65-7		Insoluble Sinker
Mercuric cyanide	Cyanides and nitriles Heavy metals	592-04-1	Poison	Soluble
Mercuric nitrate	Nitrates and nitrites Heavy metals	10045-94-0	Oxidizer	Soluble
Mercuric sulfate	Sulfates Heavy metals	7783-35-9	Poison	Insoluble Sinker
Mercuric thiocyanate	Cyanates Heavy metals	592-85-8	Poison	Insoluble Sinker
Mercurous nitrate	Nitrates and nitrites Heavy metals	7782-86-7 10415-75-5	Oxidizer	Soluble
Mercury	Heavy metals	7439-97-6		Insoluble Sinker
Mercury fulminate	Cyanides and nitriles Heavy metals	628-86-4	Explosive	Insoluble Sinker
Methacrylonitrile	Cyanides and nitriles	126-98-7	Flammable w/toxic products Polymerizable	Soluble
Methanol	Alcohols and glycols	67-56-1	Flammable	Soluble
Methapyriline (Note 1)	Amines, aryl	91-80-5		Soluble
Methomyl	Amides, anilides, and imides	16752-77-5		Soluble
Methoxychlor	Aromatics, halogenated	72-43-5		Insoluble Sinker

TABLE 1. RELEASES IN WATER

Hazardous Substance	Chemical Class	CAS No.	Hazard(s), in Addition to Toxicity	Behavior in Water
Methyl bromide	Halides, alkyl	74-83-9	Combustible w/toxic products Poison	Insoluble Sinker
Methyl chlorocarbonate	Ethers Aliphatics, halogenated	79-22-1	Flammable w/toxic products	Soluble
Methyl chloroform	Halides, alkyl	71-55-6	Combustible w/toxic products	Insoluble Sinker
Methyl ethyl ketone	Ketones	78-93-3	Flammable	Soluble
Methyl hydrazine	Hydrazines and hydrazides	60-34-4	Flammable w/toxic products Poison	Soluble
Methyl iodide	Halides, alkyl	74-88-4	Potential carcinogen	Soluble
Methyl isobutyl ketone	Ketones	108-10-1	Flammable	Soluble
Methyl isocyanate	Cyanates	624-83-9	Flammable w/toxic products	Exothermic Reaction
Methyl methacrylate	Esters, Olefins	80-62-6	Flammable Polymerizable	Soluble
Methyl parathion	Organophosphates	298-00-0	Combustible w/toxic products Poison	Insoluble Sinker
3-Methylcholanthrene	Aromatics	56-49-5	Potential carcinogen	Insoluble Sinker
Methylene bromide	Halides, alkyl	74-95-3		Soluble
Methylene chloride	Halides, alkyl	75-09-2	Combustible w/toxic products	Soluble
4,4'-Methylenebis(2-chloro-aniline) (Note 2)	Aromatics, halogenated Amines, aryl	101-14-4	Potential carcinogen	Insoluble Sinker
Methylthiouracil (Note 1)	Amines, alkyl	56-04-2	Potential carcinogen	Insoluble Sinker

TABLE 1. RELEASES IN WATER

Hazardous Substance	Chemical Class	CAS No.	Hazard(s), in Addition to Toxicity	Behavior in Water
N-Methyl-N'-nitro-N-nitrosoguanidine (Notes 1,2)	Nitro compounds Nitroso compounds	70-25-7	Flammable w/toxic products Potential carcinogen	Insoluble Sinker
Mevinphos	Organophosphates	7786-34-7	Combustible w/toxic products Poison	Soluble
Mexacarbate	Esters	315-18-4	Combustible w/toxic products Poison	Insoluble Sinker
Mitomycin C (Note 1)	(See mitomycin)	50-07-7	Potential carcinogen	Soluble
Monoethylamine	Amines, alkyl	75-04-7	Flammable w/toxic products Corrosive	Soluble
Naled	Aliphatics, halogenated Organophosphates	300-76-5	Combustible w/toxic products	Insoluble Sinker
Naphthalene	Aromatics	91-20-3	Combustible	Insoluble Sinker
Naphthenic acid	Acidic compounds, organic	1338-24-5	Combustible	Insoluble Floater
1,4-Naphthoquinone	Aromatics	130-15-4		Soluble
2-Naphthylamine	Amines, aryl	91-59-8	Potential carcinogen	Soluble
1-Naphthylamine	Amines, aryl	134-32-7		Insoluble Sinker
alpha-Naphthylthiourea	Ureas	86-88-4		Insoluble Sinker
Nickel	Heavy metals	7440-02-0	Flammable w/toxic products Potential carcinogen	Insoluble Sinker
Nickel ammonium sulfate	Sulfates Heavy metals	15699-18-0		Soluble

TABLE 1. RELEASES IN WATER

Hazardous Substance	Chemical Class	CAS No.	Hazard(s), in Addition to Toxicity	Behavior in Water
Nickel carbonyl	Organometallics Heavy metals	13463-39-3	Flammable w/toxic products Reactive Potential carcinogen	Insoluble Sinker
Nickel chloride	Halides, inorganic Heavy metals	7718-54-9 37211-05-5		Soluble
Nickel cyanide	Cyanides and nitriles Heavy metals	557-19-7	Poison	Insoluble Sinker
Nickel hydroxide	Basic compounds Heavy metals	12054-48-7		Insoluble Sinker
Nickel nitrate	Nitrates and nitrites Heavy metals	14216-75-2	Oxidizer	Soluble
Nickel sulfate	Sulfates Heavy metals	7786-81-4		Soluble
Nicotine and salts	Amines, aryl	54-11-5	Combustible w/toxic products Poison	Soluble
Nitric acid	Acidic compounds, inorganic	7697-37-2	Corrosive Reactive Oxidizer	Soluble
p-Nitroaniline	Nitro compounds Amines, aryl	100-01-6	Poison	Insoluble Sinker
Nitrobenzene	Nitro compounds Aromatics	98-95-3	Combustible w/toxic products Poison	Insoluble Sinker
Nitrogen dioxide	Acidic compounds, inorganic	10102-44-0	Corrosive Oxidizer Poison	Decomposes (Sinker)
Nitroglycerine	Nitro compounds	55-63-0	Explosive Flammable w/toxic products	Insoluble Sinker
4-Nitrophenol	Nitro compounds Phenols and cresols	100-02-7	Combustible w/toxic products	Soluble
2-Nitrophenol	Nitro compounds Phenols and cresols	88-75-5	Combustible w/toxic products	Insoluble Sinker

TABLE 1. RELEASES IN WATER

Hazardous Substance	Chemical Class	CAS No.	Hazard(s), in Addition to Toxicity	Behavior in Water
Nitrophenol (mixed)	Nitro compounds Phenols and cresols	25154-55-6	Combustible w/toxic products	Soluble
2-Nitropropane	Nitro compounds	79-46-9	Flammable w/toxic products	Soluble
N-Nitrosodi-n-butyl-amine	Nitroso compounds	924-16-3	Combustible w/toxic products Potential carcinogen	Soluble
N-Nitrosodiethanolamine	Nitroso compounds	1116-54-7	Combustible w/toxic products Potential carcinogen	Soluble
N-Nitrosodiethylamine	Nitroso compounds	55-18-5	Combustible w/toxic products Potential carcinogen	Soluble
N-Nitrosodimethylamine	Nitroso compounds	62-75-9	Flammable w/toxic products Potential carcinogen	Soluble
N-Nitrosodiphenyl-amine	Nitroso compounds Amines, aryl	86-30-6	Potential carcinogen	Insoluble Sinker
N-Nitrosodi-n-propyl-amine	Nitroso compounds	621-64-7	Combustible w/toxic products Potential carcinogen	Insoluble Floater
N-Nitroso-N-ethylurea (Note 2)	Nitroso compounds	759-73-9	Potential carcinogen	Soluble
N-Nitroso-N-methylurea (Note 1)	Nitroso compounds	684-93-5	Combustible w/toxic products Potential carcinogen	Soluble
N-Nitroso-N-methyl-urethane (Note 2)	Nitroso compounds	615-53-2	Combustible w/toxic products Potential carcinogen	Soluble
N-Nitrosomethylvinyl-amine	Nitroso compounds	4549-40-0	Flammable w/toxic products Potential carcinogen	Soluble

TABLE 1. RELEASES IN WATER

Hazardous Substance	Chemical Class	CAS No.	Hazard(s), in Addition to Toxicity	Behavior in Water
N-Nitrosopiperidine	Nitroso compounds	100-75-4	Potential carcinogen	Soluble
N-Nitrosopyrrolidine	Nitroso compounds	930-55-2	Combustible w/toxic products Potential carcinogen	Soluble
Nitrotoluene	Nitro compounds Aromatics	1321-12-6	Combustible w/toxic products	Insoluble Sinker
5-Nitro-o-toluidine (Note 2)	Nitro compounds Aromatics	99-55-8		Insoluble Sinker
Octamethylpyrophosphor-amide	Amides, anilides, and imides Organophosphates	152-16-9		Soluble
Osmium tetroxide	Oxides Heavy metals	20816-12-0		Soluble
Paraformaldehyde	Aldehydes	30525-89-4	Combustible	Soluble
Paraldehyde	Aldehydes	123-63-7	Flammable	Soluble
Parathion	Organophosphates	56-38-2	Poison	Insoluble Sinker
Pentachlorobenzene	Aromatics, halogenated	608-93-5	Combustible w/toxic products	Insoluble Sinker
Pentachloroethane	Aliphatics, halogenated	76-01-7	Combustible w/toxic products	Insoluble Sinker
Pentachloronitr-benzene	Nitro compounds Aromatics, halogenated	82-68-8		Insoluble Sinker
Pentachlorophenol	Phenols and cresols Aromatics, halogenated	87-86-5		Insoluble Sinker
1,3-Pentadiene	Olephins	504-60-9	Flammable	Insoluble Floater
Phenacetin	Aromatics Amides, anilides, and imides	62-44-2	Potential carcinogen	Insoluble Sinker

TABLE 1. RELEASES IN WATER

Hazardous Substance	Chemical Class	CAS No.	Hazard(s), in Addition to Toxicity	Behavior in Water
Phenanthrene	Aromatics	85-01-8		Insoluble Sinker
Phenol	Phenols and cresols	108-95-2	Combustible Corrosive Poison	Soluble
Phenyl dichloroarsine	Aromatics, halogenated Heavy metals	696-28-6	Poison	Insoluble Sinker
Phenylmercuric acetate	Organometallics Heavy metals	62-38-4	Combustible w/toxic products	Insoluble Sinker
N-Phenylthiourea	Ureas Aromatics	103-85-5		Insoluble Sinker
Phorate	Organophosphates	298-02-2		Insoluble Sinker
Phosphoric acid	Acids, inorganic	7664-38-2	Corrosive	Soluble
Phosphorus	Phosphorous and compounds	7723-14-0	Flammable w/toxic products Poison	Insoluble Sinker
Phosphorus oxychloride	Phosphorous and compounds Halides, inorganic	10025-87-3	Corrosive Reactive	Reacts violently to give off HCl
Phosphorus pentasulfide	Phosphorous and compounds Sulfides and mercaptans	1314-80-3	Flammable w/toxic products Reactive	Forms hydrogen sulfide on contact
Phosphorus trichloride	Phosphorous and compounds Halides, inorganic	7719-12-2	Corrosive Reactive	Reacts violently to give off HCl
Phthalic anhydride	Aromatics	85-44-9	Corrosive	Insoluble Sinker
2-Picoline	Amines, aryl	109-06-8	Combustible w/toxic products	Soluble

TABLE 1. RELEASES IN WATER

Hazardous Substance	Chemical Class	CAS No.	Hazard(s), in Addition to Toxicity	Behavior in Water
Polychlorinated biphenyls	Aromatics, halogenated	1336-36-3	Potential carcinogen	Insoluble Sinker
Potassium arsenate	Heavy metals	7784-41-0	Poison	Soluble
Potassium arsenite	Heavy metals	10124-50-2	Poison	Soluble
Potassium bichromate	Chromates	7778-50-9	Corrosive	Soluble
Potassium chromate	Chromates	7789-00-6		Soluble
Potassium cyanide	Cyanides and nitriles	151-50-8	Poison	Produces cyanide ion on contact
Potassium hydroxide	Basic compounds	1310-58-3	Corrosive	Generates heat on contact
Potassium permanganate	Basic compounds	7722-64-7	Corrosive Oxidizer	Soluble
Potassium silver cyanid	Cyanides and nitriles Heavy metals	506-61-6		Soluble
Pronamide (Note 2)	Aromatics, halogenated Amides, anilides, and imides	23950-58-5		Insoluble Sinker
1,3-Propane sultone	Sulfones, sulfoxides, and sulfonates	1120-71-4	Potential carcinogen	Soluble
Propargite	Sulfites Aromatics	2312-35-8	Flammable w/toxic products	Insoluble Sinker
Propargyl alcohol	Alcohols and glycols	107-19-7	Flammable Poison	Soluble
Propionic acid	Acidic compounds, organic	79-09-4	Combustible	Soluble
Propionic anhydride	Acidic compounds, organic	123-62-6	Combustible	Decomposes (Sinker)
n-Propylamine	Amines, alkyl	107-10-8	Flammable w/toxic products	Soluble

TABLE 1. RELEASES IN WATER

Hazardous Substance	Chemical Class	CAS No.	Hazard(s), in Addition to Toxicity	Behavior in Water
Propylene dichloride	Aliphatics, halogenated	78-87-5	Flammable w/toxic products	Insoluble Sinker
Propylene oxide	Oxides, alkylene	75-56-9	Flammable Corrosive	Soluble
1,2-Propylenimine	Amines, alkyl	75-55-8	Flammable w/toxic products Potential carcinogen	Soluble
Pyrene	Aromatics	129-00-0	Combustible w/toxic products	Insoluble Sinker
Pyrethrins	Acidic compounds, organic	121-21-1 121-29-9	Combustible w/toxic products	Insoluble Sinker
4-Pyridinamine (Note 2)	Amines, alkyl	504-24-5	Combustible w/toxic products	Soluble
Pyridine	Amines, aryl	110-86-1	Flammable w/toxic products	Soluble
Quinoline	Amines, aryl	91-22-5	Combustible	Soluble
Reserpine	Aromatics	50-55-5		Insoluble Sinker
Resorcinol	Aromatics	108-46-3	Combustible	Soluble
Saccharin	Sulfones, sulfoxides, and sulfonates Aromatics	81-07-2	Potential carcinogen	Insoluble Floater
Safrole	Aromatics Ethers	94-59-7	Combustible Potential carcinogen	Insoluble Sinker
Selenious acid	Acidic compounds, inorganic Heavy metals	7783-00-8		Soluble
Selenium	Heavy metals	7782-49-2	Combustible w/toxic products	Insoluble Sinker
Selenium disulfide	Sulfides and mercaptans Heavy metals	7488-56-4	Reactive	Forms hydrogen sulfide on contact

TABLE 1. RELEASES IN WATER

Hazardous Substance	Chemical Class	CAS No.	Hazard(s), in Addition to Toxicity	Behavior in Water
Selenium oxide	Oxides Heavy metals	7446-08-4	Poison	Soluble
Selenourea (Note 2)	Ureas Heavy metals	630-10-4		Soluble
Silver	Heavy metals	7440-22-4		Insoluble Sinker
Silver cyanide	Cyanides and nitriles Heavy metals	506-64-9	Poison	Insoluble Sinker
Silver nitrate	Nitrates and nitrites Heavy metals	7761-88-8	Oxidizer	Soluble
Sodium	Alkali metals	7440-23-5	Flammable Corrosive Reactive	Inflames on contact
Sodium arsenate	Heavy metals	7631-89-2	Poison	Soluble
Sodium arsenite	Heavy metals	7784-46-5	Poison	Soluble
Sodium azide	Azo compounds	26628-22-8	Combustible w/toxic products Poison Explosive	Soluble
Sodium bichromate	Chromates	10588-01-9	Corrosive Oxidizer	Soluble
Sodium bifluoride	Halides, inorganic	1333-83-1	Corrosive	Soluble
Sodium bisulfite	Sulfites	7631-90-5		Soluble
Sodium chromate	Chromates	7775-11-3		Soluble
Sodium cyanide	Cyanides and nitriles	143-33-9	Poison	Soluble
Sodium dodecylbenzene sulfonate (Note 2)	Sulfones, sulfoxides, and sulfonates	25155-30-0		Soluble

TABLE 1. RELEASES IN WATER

Hazardous Substance	Chemical Class	CAS No.	Hazard(s), in Addition to Toxicity	Behavior in Water
Sodium fluoracetate	Organometallics	62-74-8		Soluble
Sodium fluoride	Halides, inorganic	7681-49-4		Reacts to form hydrogen fluoride
Sodium hydrosulfide	Sulfides and mercaptans	16721-80-5	Flammable w/toxic products Reactive	Soluble
Sodium hydroxide	Basic compounds	1310-73-2	Corrosive Reactive	Generates heat on contact
Sodium hypochlorite	Halides, inorganic	10022-70-5 7681-52-9		Soluble
Sodium methylate	Organometallics	124-41-4	Flammable w/toxic products Reactive	Soluble
Sodium nitrite	Nitrates and nitrites	7632-00-0	Oxidizer	Soluble
Sodium phosphate, dibasic	Phosphates and phosphonates	7558-79-4 10039-32-4 10028-24-7 10140-65-5		Soluble
Sodium phosphate, tribasic	Phosphates and phosphonates	7601-54-9 7785-84-4 10101-89-0 10361-89-4 7758-29-4 10124-56-8		Soluble
Sodium selenite	Heavy metals	10102-18-8 7782-82-3	Poison	Soluble
Streptozotocin (Note 1)	(See streptozotocin)	18883-66-4	Potential carcinogen	Soluble
Strontium chromate	Chromates Heavy metals	7789-06-2		Insoluble Sinker
Strontium sulfide	Sulfides and mercaptans Heavy metals	1314-96-1	Reactive	Reacts to give toxic hydrogen sulfide

TABLE 1. RELEASES IN WATER

Hazardous Substance	Chemical Class	CAS No.	Hazard(s), in Addition to Toxicity	Behavior in Water
Strychnine and salts	(See strychnine and salts)	57-24-9	Poison	Insoluble Sinker
Styrene	Aromatics, Olefins	100-42-5	Flammable Polymerizable	Insoluble Floater
Sulfur monochloride	Halides, inorganic	12771-08-3	Corrosive Reactive	Decomposes (Sinker)
Sulfuric acid	Acidic compounds, inorganic	7664-93-9	Corrosive Reactive	Soluble with evolution of heat
sym-Trinitrobenzene	Nitro compounds Aromatics	99-35-4	Flammable w/toxic products Explosive	Insoluble Sinker
2,4,5-T acid	Aromatics, halogenated	93-73-5		Insoluble Sinker
2,4,5-T amines	Amines, aryl Aromatics, halogenated	2008-46-0	Combustible w/toxic products	Insoluble Sinker
2,4,5-T esters	Esters Aromatics, halogenated	N.A.		Insoluble Sinker
2,4,5-T salts	Aromatics, halogenated	13560-99-1	Combustible w/toxic products	Soluble
1,2,4,5-Tetrachlorobenzene	Aromatics, halogenated	95-94-3		Insoluble Sinker
1,1,1,2-Tetrachloroethane	Aliphatics, halogenated	630-20-6		Insoluble Sinker
1,1,2,2-Tetrachloroethane	Aliphatics, halogenated	79-34-5		Insoluble Sinker
2,3,4,6-Tetrachlorophenol	Aromatics, halogenated Phenols and cresols	58-90-2	Combustible w/toxic products	Insoluble Sinker
2,3,7,8-Tetrachlorodibenzo-p-dioxin	Aromatics, halogenated	1746-01-6	Combustible w/toxic products Potential carcinogen	Insoluble Sinker
Tetrachloroethylene	Aliphatics, halogenated Olefins	127-18-4		Insoluble Sinker

TABLE 1. RELEASES IN WATER

Hazardous Substance	Chemical Class	CAS No.	Hazard(s), in Addition to Toxicity	Behavior in Water
Tetraethyldithiopyro-phosphate	Organophosphates	3689-24-5	Poison	Insoluble Sinker
Tetraethyl lead	Organometallics Heavy metals	78-00-2	Combustible w/toxic products Poison	Insoluble Sinker
Tetraethyl pyrophos-phate	Organophosphates	107-49-3	Combustible w/toxic products Poison	Soluble
Tetrahydrofuran	Ethers	109-99-9	Flammable	Soluble
Tetranitromethane	Nitro compounds	509-14-8	Explosive Oxidizer	Insoluble Sinker
Thallium	Heavy metals	7440-28-0		Insoluble Sinker
Thallium(I) acetate	Organometallics	563-68-8		Soluble
Thallium(I) carbonate	Organometallics Heavy metals	6533-73-9		Soluble
Thallium (I) chloride	Halides, inorganic Heavy metals	7791-12-0	Poison	Insoluble Sinker
Thallium(I) nitrate	Nitrates and nitrites Heavy metals	10102-45-1	Oxidizer	Soluble
Thallium (III) oxide	Oxides Heavy metals	1314-32-5	Oxidizer	Insoluble Sinker
Thallium (I) selenide	Heavy metals	12039-52-0		Insoluble Sinker
Thallium(I) sulfate	Sulfates Heavy metals	7446-18-6	Poison	Soluble
1-(o-Chlorophenyl)-thiourea (Note 2)	Aromatics, halogenated Ureas	5344-82-1		Soluble
Thioacetamide (Note 2)	Amides, anilides, and imides	62-55-5	Potential carcinogen	Soluble

TABLE 1. RELEASES IN WATER

Hazardous Substance	Chemical Class	CAS No.	Hazard(s), in Addition to Toxicity	Behavior in Water
Thiofanox (Note 2)	Sulfides and mercaptans Amides, anilines, and imides	39196-18-4		Insoluble Sinker
Thiophenol	Phenols and cresols	108-98-5		Insoluble Sinker
Thiosemicarbazide (Note 1)	Azo compounds Ureas	79-19-6		Soluble
Thiourea	Ureas	62-56-6	Potential carcinogen	Soluble
Thiram	Sulfides and mercaptans	137-26-8	Combustible w/toxic products	Insoluble Sinker
Toluene	Aromatics	108-88-3	Flammable	Insoluble Floater
Toluene diisocyanate	Cyanates Aromatics	584-84-9	Combustible w/toxic products Poison	Decomposes (Sinker)
Toluenediamine	Amines, aryl	95-80-7	Potential carcinogen	Soluble
o-Toluidine hydrochloride	Aromatics, halogenated Amines, aryl	636-21-5	Potential carcinogen	Soluble
Toxaphene	Aliphatics, halogenated	8001-35-2	Combustible w/toxic products Potential carcinogen	Insoluble Sinker
2,4,5-TP acid	Acidic compounds, organic Aromatics, halogenated	93-72-1	Combustible w/toxic products	Insoluble Sinker
2,4,5-TP acid esters	Esters, Aromatics, halogenated	N.A.	Combustible w/toxic products	Insoluble Sinker
Tris(2,3-dibromopropyl) phosphate (Note 2)	Phosphates and phosphonates	126-72-7	Potential carcinogen	Insoluble Sinker
Trichlorfon	Aliphatics, halogenated Organophosphates	52-68-6		Soluble

TABLE 1. RELEASES IN WATER

Hazardous Substance	Chemical Class	CAS No.	Hazard(s), in Addition to Toxicity	Behavior in Water
1,2,4-Trichlorobenzene	Aromatics, halogenated	120-82-1	Combustible w/toxic products	Insoluble Sinker
1,1,2-Trichloroethane	Aliphatics, halogenated	79-00-5	Combustible w/toxic products	Insoluble Sinker
Trichloroethylene	Aliphatics, halogenated	79-01-6	Flammable w/toxic products	Insoluble Sinker
Trichloromethanesul-fenyl chloride (Note 2)	Aliphatics, halogenated	594-42-3	Poison	Insoluble Sinker
Trichloromonofluoro-methane	Aliphatics, halogenated	75-69-4		Insoluble Sinker
2,4,5-Trichlorophenol	Phenols and cresols Aromatics, halogenated	95-95-4	Combustible w/toxic products	Insoluble Sinker
2,4,6-Trichlorophenol	Phenols and cresols Aromatics, halogenated	88-06-2	Combustible w/toxic products	Insoluble Sinker
Trichlorophenol	Phenols and cresols Aromatics, halogenated	25167-82-2		Insoluble Sinker
Triethanolamine dode-cylbenzene sulfonate	Sulfones, sulfoxides, and sulfonates	27323-41-7		Soluble
Triethylamine	Amines, alkyl	121-44-8	Flammable w/toxic products Corrosive	Soluble
Trypan blue (Note 2)	Azo compounds	72-57-1	Potential carcinogen	Soluble
Uracil mustard	Aliphatics, halogenated Amines, alkyl	66-75-1	Potential carcinogen	Insoluble Sinker
Uranyl acetate	Organometallics Heavy metals	541-09-3	Radioactive	Soluble
Uranyl nitrate	Nitrates and nitrites Heavy metals	10102-06-4 36478-76-9	Radioactive Oxidizer	Soluble
Vanadium pentoxide	Oxides Heavy metals	1314-62-1		Insoluble Sinker
Vanadyl sulfate	Sulfates Heavy metals	27774-13-6		Soluble

TABLE 1. RELEASES IN WATER

Hazardous Substance	Chemical Class	CAS No.	Hazard(s), in Addition to Toxicity	Behavior in Water
Vinyl acetate	Esters, Olefins	108-05-4	Flammable Polymerizable	Soluble
Vinylidene chloride	Aliphatics, halogenated	75-35-4	Flammable w/toxic products Polymerizable Potential carcinogen	Insoluble Sinker
Warfarin	Aromatics	81-81-2		Insoluble Sinker
Xylene	Aromatics	1330-20-7	Flammable	Insoluble Floater
Xylenol	Phenols and cresols	1300-71-6	Combustible	Soluble
Zinc	Heavy metals	7440-66-6	Combustible	Insoluble Sinker
Zinc acetate	Organometallics Heavy metals	557-34-6		Soluble
Zinc ammonium chloride	Halides, inorganic Heavy metals	52628-25-8 14639-97-5 14639-98-6		Soluble
Zinc borate	Heavy metals	1332-07-6		Soluble
Zinc bromide	Halides, inorganic Heavy metals	7699-45-8		Soluble
Zinc carbonate	Organometallics Heavy metals	3486-35-9		Insoluble Sinker
Zinc chloride	Halides, inorganic Heavy metals	7646-85-7		Soluble
Zinc cyanide	Cyanides and nitriles Heavy metals	557-21-1	Poison	Insoluble Sinker
Zinc fluoride	Halides, inorganic Heavy metals	7783-49-5		Insoluble Sinker

TABLE 1. RELEASES IN WATER

Hazardous Substance	Chemical Class	CAS No.	Hazard(s), in Addition to Toxicity	Behavior in Water
Zinc formate	Organometallics Heavy metals	557-41-5		Soluble
Zinc hydrosulfite	Sulfites Heavy metals	7779-86-4		Soluble
Zinc nitrate	Nitrates and nitrites Heavy metals	7779-88-6	Oxidizer	Soluble
Zinc phenolsulfonate	Phenols and cresols Heavy metals	127-82-2	Combustible w/toxic products	Soluble
Zinc phosphide	Phosphorous and compounds Heavy metals	1314-84-7	Flammable w/toxic products Reactive Poison	Insoluble Sinker
Zinc silicofluoride	Halides, inorganic Heavy metals	16871-71-9		Soluble
Zinc sulfate	Sulfates Heavy metals	7733-02-0		Soluble
Zirconium nitrate	Nitrates and nitrites Heavy metals	13746-89-9	Oxidizer	Soluble
Zirconium potassium fluoride	Halides, inorganic· Heavy metals	16923-95-8		Soluble
Zirconium sulfate	Sulfates Heavy metals	14644-61-2		Soluble
Zirconium tetrachloride	Halides, inorganic Heavy metals	10026-11-6	Corrosive Reactive	Soluble

Note 1: Specific gravity calculated according to Grains Method given in the Handbook of Chemical Property Estimation Methods.

Note 2: Specific gravity and/or solubility estimated based on active chemical groups and physical characteristics of structurally similar compounds.

LEGEND: N/A = Not available

TABLE 2. LIQUIDS RELEASED ON LAND

Hazardous Substance	Chemical Class	CAS No.	Hazard(s), in Addition to Toxicity
Acetaldehyde	Aldehydes	75-07-0	Flammable Polymerizable
Acetic acid	Acidic compounds, organic	64-19-7	Combustible Corrosive
Acetic anhydride	Acidic compounds, organic	108-24-7	Combustible Corrosive
Acetone	Ketones	67-64-1	Flammable
Acetone cyanohydrin	Cyanides and nitriles	75-86-5	Combustible w/toxic products Poison
Acetonitrile	Cyanides and nitriles	75-05-8	Flammable w/toxic products
Acetophenone	Ketones	98-86-2	Combustible
Acetyl bromide	Aliphatics, halogenated	506-96-7	Flammable w/toxic products Corrosive Reactive
Acetyl chloride	Aliphatics, halogenated	75-36-5	Flammable w/toxic products Corrosive Reactive
Acrolein	Aldehydes, Olefins	107-02-8	Flammable Polymerizable Poison
Acrylic acid	Acidic compounds, organic, Olefins	79-10-7	Combustible Corrosive Polymerizable
Acrylonitrile	Cyanides and nitriles	107-13-1	Flammable w/toxic products Polymerizable Potential carcinogen Poison
Allyl alcohol	Alcohols and glycols, Olefins	107-18-6	Flammable Poison
Allyl chloride	Halides, alkyl, Olefins	107-05-1	Flammable w/toxic products

TABLE 2. LIQUIDS RELEASED ON LAND

Hazardous Substance	Chemical Class	CAS No.	Hazard(s), in Addition to Toxicity
Ammonium hydroxide	Basic compounds	1336-21-6	Corrosive
Ammonium sulfide (in aqueous solution)	Sulfides and mercaptans	12135-76-1	Flammable w/toxic products
Amyl acetate	Esters	628-63-7	Flammable
Aniline	Amines, aryl	62-53-3	Combustible w/toxic products Poison
Antimony pentachloride	Halides, inorganic, Heavy metals	7647-18-9	Corrosive Reactive
Arsenic trichloride	Halides, inorganic, Heavy metals	7784-34-1	Corrosive Reactive Poison
Benzal chloride	Aromatics, halogenated	98-87-3	
Benzene	Aromatics	71-43-2	Flammable Potential carcinogen
Benzenesulfonyl chloride	Acidic compounds, organic	98-09-9	Combustible w/toxic products
Benzonitrile	Cyanides and nitriles	100-47-0	Combustible w/toxic products
Benzotrichloride	Aromatics, halogenated	98-07-7	Combustible w/toxic products Corrosive
Benzoyl chloride	Aromatics, halogenated	98-88-4	Combustible w/toxic products Corrosive
Benzyl chloride	Aromatics, halogenated	100-44-7	Combustible w/toxic products Corrosive Reactive
2,2'-Bioxirane	Epoxides	1464-53-5	Potential carcinogen
Bis(2-chloroethoxy) methane	Aliphatics, halogenated	111-91-1	
Bis(2-chloroethyl) ether	Ethers	111-44-4	Combustible w/toxic products Poison

TABLE 2. LIQUIDS RELEASED ON LAND

Hazardous Substance	Chemical Class	CAS No.	Hazard(s), in Addition to Toxicity
Bis(2-chloroisopropyl)-ether	Ethers, Aliphatics, halogenated	108-60-1	Combustible w/toxic products
Bis(chloromethyl) ether	Ethers	542-88-1	Combustible w/toxic products Potential carcinogen
Bis(2-ethylhexyl) phthalate	Esters	117-81-7	
Bromoacetone (Note 2)	Ketones	598-31-2	Poison
Bromoform	Halides, alkyl	75-25-2	
4-Bromophenyl phenyl ether	Ethers, Aromatics, halogenated	101-55-3	Combustible w/toxic products
1-Butanol	Alcohols and glycols	71-36-3	Flammable
2-Butanone peroxide (Note 2)	Peroxides	1338-23-4	Explosive Oxidizer Combustible
Butyl acetate	Esters	123-86-4	Flammable
Butyl benzyl phthalate	Esters	85-68-7	
Butylamine	Amines, alkyl	109-73-9	Flammable w/toxic products
Butyric acid	Acidic compounds, organic	107-92-6	Combustible
Carbon disulfide	Sulfides and mercaptans	75-15-0	Flammable w/toxic products
Carbon tetrachloride	Halides, alkyl	56-23-5	Potential carcinogen
Chloral	Aldehydes, Aliphatics, halogenated	75-87-6	Combustible w/toxic products Corrosive Potential carcinogen
Chloroacetaldehyde	Aldehydes, Aliphatics, halogenated	107-20-0	Combustible w/toxic products Polymerizable
Chlorobenzene	Aromatics, halogenated	108-90-7	Flammable w/toxic products

TABLE 2. LIQUIDS RELEASED ON LAND

Hazardous Substance	Chemical Class	CAS No.	Hazard(s), in Addition to Toxicity
Chlorodibromomethane	Aliphatics, halogenated	124-48-1	
Chloroethane	Aliphatics, halogenated	75-00-3	Flammable w/toxic products
2-Chloroethyl vinyl ether	Ethers, Aliphatics, halogenated	110-75-8	Flammable w/toxic products
Chloroform	Halides, alkyl	67-66-3	Potential carcinogen
Chloromethyl methyl ether	Ethers, Aliphatics, halogenated	107-30-2	Flammable w/toxic products Potential carcinogen Poison
2-Chlorophenol	Phenols and cresols, Aromatics, halogenated	95-57-8	Combustible w/toxic products
4-Chlorophenyl phenyl ether	Ethers, Aromatics, halogenated	7005-72-3	Combustible w/toxic products
3-Chloropropionitrile	Cyanides and nitriles	542-76-7	Combustible w/toxic products
Chlorosulfonic acid	Acidic compounds, inorganic	7790-94-5	Corrosive Reactive
Creosote	Phenols and cresols, Aromatics	8001-58-9	Combustible Potential carcinogen
Cresol	Phenols and cresols, Aromatics	1319-77-3	Combustible
Crotonaldehyde	Aldehydes, Olefins	4170-30-3 123-73-9	Flammable
Cumene	Aromatics	98-82-8	Combustible
Cyanogen chloride	Cyanides and nitriles	506-77-4	Poison
Cyclohexane	Aliphatics	110-82-7	Flammable
Cyclohexanone	Ketones	108-94-1	Combustible
2,4-D esters	Esters	94-11-1	Combustible w/toxic products
Diazinon	Organophosphates	333-41-5	Combustible w/toxic products

TABLE 2. LIQUIDS RELEASED ON LAND

Hazardous Substance	Chemical Class	CAS No.	Hazard(s), in Addition to Toxicity
1,2-Dibromo-3-chloropropane (Note 2)	Aliphatics, halogenated	96-12-8	Combustible w/toxic products Potential carcinogen
Di-n-butylphthalate	Esters	84-74-2	
Dichlorobenzene (all isomers)	Aromatics, halogenated	25321-22-6	Combustible w/toxic products
o-Dichlorobenzene	Aromatics, halogenated	95-50-1	Combustible w/toxic products
m-Dichlorobenzene	Aromatics, halogenated	541-73-1	Combustible w/toxic products
Dichlorobromomethane	Aliphatics, halogenated	75-27-4	Combustible w/toxic products
1,4-Dichloro-2-butene	Aliphatics, halogenated	764-41-0	Flammable w/toxic products
1,1-Dichloroethane	Aliphatics, halogenated	75-34-3	Flammable w/toxic products
1,2-trans-Dichloroethylene	Aliphatics, halogenated	156-60-5	Flammable w/toxic products
Dichloropropane (all isomers)	Aliphatics, halogenated	26638-19-7	Flammable w/toxic products
1,3-Dichloropropene	Aliphatics, halogenated	542-75-6	Flammable w/toxic products
Dichloropropene (all isomers)	Aliphatics, halogenated	26952-23-8	Flammable w/toxic products
Dichloropropene-Dichloropropane mixture	Aliphatics, halogenated	8003-19-8	Flammable w/toxic products
2,2-Dichloropropionic acid	Acidic compounds, organic, Aliphatics, halogenated	75-99-0	Corrosive
Dichlorvos	Organophosphates	62-73-7	Poison
Diethylamine	Amines, alkyl	109-89-7	Flammable w/toxic products Corrosive

TABLE 2. LIQUIDS RELEASED ON LAND

Hazardous Substance	Chemical Class	CAS No.	Hazard(s), in Addition to Toxicity
Diethylarsine (Note 2)	Organometallics, Heavy metals	692-42-2	Flammable w/toxic products
N,N'-Diethylhydrazine	Hydrazines and hydrazides	1615-80-1	Combustible w/toxic products Potential carcinogen
O,O-Diethyl-S-methyl dithiophosphate (Note 2)	Organophosphates	3288-58-2	
Diethyl-p-nitrophenyl phosphate	Organophosphates, Nitro compounds	311-45-5	
Diethyl phthalate	Esters, Aromatics	84-66-2	
O,O-Diethyl O-pyrazinyl phosphorothioate	Organophosphates	297-97-2	
Dihydrosafrole (Note 2)	Aromatics, Ethers	94-58-6	Potential carcinogen
Diisopropyl fluorophosphate	Organophosphates	55-91-4	Corrosive
alpha,alpha-Dimethylbenzylhydroperoxide	Peroxides, Aromatics	80-15-9	Explosive Combustible
Dimethyl carbamoyl chloride (Note 2)	Halides, alkyl	79-44-7	Combustible w/toxic products Reactive Potential carcinogen
alpha,alpha-Dimethylphenethylamine (Note 1)	Amines, aryl	122-09-8	
Dimethyl phthalate	Esters	131-11-3	Combustible
Dimethyl sulfate	Sulfates	77-78-1	Combustible w/toxic products Corrosive Potential carcinogen
1,1-Dimethylhydrazine	Hydrazines and hydrazides	57-14-7	Flammable w/toxic products Corrosive Potential carcinogen Poison

TABLE 2. LIQUIDS RELEASED ON LAND

Hazardous Substance	Chemical Class	CAS No.	Hazard(s), in Addition to Toxicity
1,2-Dimethylhydrazine	Hydrazines and hydrazides	540-73-8	Flammable w/toxic products Corrosive Potential carcinogen
Di-n-octylphthalate	Esters	117-84-0	
1,4-Dioxane	Ethers	123-91-1	Flammable Potential carcinogen
Dipropylamine	Amines, alkyl	142-84-7	Flammable w/toxic products
Dodecylbenzenesulfonic acid	Acids, organic, Aromatics	27176-87-0	Combustible w/toxic products
Endrin aldehyde (Note 2)	Aldehydes	930-55-2	
Epichlorohydrin	Epoxides, Aliphatics, halogenated	106-89-8	Flammable w/toxic products
Ethion	Organophosphates	563-12-2	Poison
Ethyl acetate	Esters	141-78-6	Flammable
Ethyl acrylate	Esters, Olefins	140-88-5	Flammable Polymerizable
Ethyl cyanide	Cyanides and nitriles	107-12-0	Flammable w/toxic products
Ethyl-4,4'-dichlorobenzilate (Note 1)	Esters, Aliphatics, halogenated	510-15-6	
Ethyl ether	Ethers	60-29-7	Explosive, upon standing Flammable
Ethyl methacrylate	Esters, Olefins	97-63-2	Flammable Polymerizable
Ethyl methanesulfonate (Note 2)	Esters	62-50-0	
Ethylbenzene	Aromatics	100-41-4	Flammable
Ethylene dibromide	Aliphatics, halogenated	106-93-4	Potential carcinogen
Ethylene dichloride	Aliphatics, halogenated	107-06-2	Flammable w/toxic products Potential carcinogen

TABLE 2. LIQUIDS RELEASED ON LAND

Hazardous Substance	Chemical Class	CAS No.	Hazard(s), in Addition to Toxicity
Ethylene oxide	Oxides, alkylene	75-21-8	Flammable Corrosive
Ethylenediamine	Amines, alkyl	107-15-3	Flammable w/toxic products Corrosive
Ethylenimine	Amines, alkyl	151-56-4	Flammable w/toxic products Poison
Formic acid	Acidic compounds, organic	64-18-6	Combustible
Furan	Ethers, Aromatics	110-00-9	Flammable
Furfural	Aldehydes, Olefins	98-01-1	Combustible
Glycidaldehyde (Note 2)	Aldehydes	765-34-4	Flammable Potential carcinogen
Hexachlorobutadiene	Aliphatics, halogenated, Olefins	87-68-3	
Hexachlorocyclo-pentadiene	Aliphatics, halogenated, Olefins	77-47-4	
Hexachloropropene (Note 2)	Aliphatics, halogenated, Olefins	1888-71-7	
Hexaethyl tetraphosphate	Organophosphates	757-58-4	Poison
Hydrazine	Hydrazines and hydrazides	302-01-2	Flammable w/toxic products Corrosive Potential carcinogen Poison
Hydrochloric acid	Acidic compounds, inorganic	7647-01-0	Corrosive Reactive
Hydrocyanic acid	Cyanides and nitriles, Acidic compounds, inorganic	74-90-8	Flammable w/toxic products Poison

TABLE 2. LIQUIDS RELEASED ON LAND

Hazardous Substance	Chemical Class	CAS No.	Hazard(s), in Addition to Toxicity
Hydrofluoric acid	Acidic compounds, inorganic	7664-39-3	Corrosive Reactive
Isobutyl alcohol	Alcohols and glycols	78-83-1	Flammable
Isophorone	Ketones	78-59-1	Combustible
Isoprene	Olefins	78-79-5	Flammable Polymerizable
Isosafrole (Note 2)	Aromatics, Ethers	120-58-1	Potential carcinogen
Lead fluoborate (in aqueous solution)	Heavy metals, Halides, inorganic	13814-96-5	
Malathion	Organophosphates	121-75-5	
Mercury	Heavy metals	7439-97-6	
Methacrylonitrile	Cyanides and nitriles	126-98-7	Flammable w/toxic products Polymerizable
Methanol	Alcohols and glycols	67-56-1	Flammable
Methapyriline (Note 1)	Amines, aryl	91-80-5	
Methyl chlorocarbonate	Esters, Aliphatics, halogenated	79-22-1	Flammable w/toxic products
Methyl chloroform	Halides, alkyl	71-55-6	Combustible w/toxic products
Methyl ethyl ketone	Ketones	78-93-3	Flammable
Methyl hydrazine	Hydrazines and hydrazides	60-34-4	Flammable w/toxic products Poison
Methyl iodide	Halides, alkyl	74-88-4	Potential carcinogen
Methyl isobutyl ketone	Ketones	108-10-1	Flammable
Methyl isocyanate	Cyanates	624-83-9	Flammable w/toxic products
Methyl methacrylate	Esters, Olefins	80-62-6	Flammable Polymerizable

TABLE 2. LIQUIDS RELEASED ON LAND

Hazardous Substance	Chemical Class	CAS No.	Hazard(s), in Addition to Toxicity
Methylene bromide	Halides, alkyl	74-95-3	
Methylene chloride	Halides, alkyl	75-09-2	Combustible w/toxic products
Mevinphos	Organophosphates	7786-34-7	Combustible w/toxic products Poison
Monoethylamine	Amines, alkyl	75-04-7	Flammable w/toxic products Corrosive
Naphthenic acid	Acidic compounds, organic	1338-24-5	Combustible
Nickel carbonyl	Organometallics, Heavy metals	13463-39-3	Flammable w/toxic products Reactive Potential carcinogen
Nicotine and salts	Amines, aryl	54-11-5	Combustible w/toxic products Poison
Nitric acid	Acidic compounds, inorganic	7697-37-2	Corrosive Reactive Oxidizer
Nitrobenzene	Nitro compounds, Aromatics	98-95-3	Combustible w/toxic products Poison
Nitrogen dioxide	Acidic compounds, inorganic	10102-44-0	Corrosive Oxidizer Poison
2-Nitropropane	Nitro compounds	79-46-9	Flammable w/toxic products
N-Nitrosodiethanolamine	Nitroso compounds	1116-54-7	Combustible w/toxic products Potential carcinogen
N-Nitrosodiethylamine	Nitroso compounds	55-18-5	Combustible w/toxic products Potential carcinogen

TABLE 2. LIQUIDS RELEASED ON LAND

Hazardous Substance	Chemical Class	CAS No.	Hazard(s), in Addition to Toxicity
N-Nitrosodimethylamine	Nitroso compounds	62-75-9	Flammable w/toxic products Potential carcinogen
N-Nitrosodi-n-propyl-amine	Nitroso compounds	621-64-7	Combustible w/toxic products Potential carcinogen
N-Nitroso-N-methylurea (Note 1)	Nitroso compounds	684-93-5	Combustible w/toxic products Potential carcinogen
N-Nitrosomethylvinyl-amine	Nitroso compounds	4549-40-0	Flammable w/toxic products Potential carcinogen
N-Nitrosopyrrolidine	Nitroso compounds,	930-55-2	Combustible w/toxic products Potential carcinogen
Octamethylpyrophosphor-amide	Amides, anilides and, imides, Organophosphates	152-16-9	
Paraldehyde	Ethers	123-63-7	Flammable
Parathion	Organophosphates	56-38-2	Poison
Pentachloroethane	Aliphatics, halogenated	76-01-7	Combustible w/toxic products
1,3-Pentadiene	Olefins	504-60-9	Flammable
Phosphoric acid	Acidic compounds, inorganic	7664-38-2	Corrosive
Phosphorus oxychloride	Phosphorous and compounds, Halides, inorganic	10025-87-3	Corrosive Reactive
Phosphorus trichloride	Phosphorous and compounds, Halides, inorganic	7719-12-2	Corrosive Reactive
2-Picoline	Amines, aryl	109-06-8	Combustible w/toxic products
Polychlorinated biphenyls	Aromatics, halogenated	1336-36-3	Potential carcinogen

TABLE 2. LIQUIDS RELEASED ON LAND

Hazardous Substance	Chemical Class	CAS No.	Hazard(s), in Addition to Toxicity
Propargite	Sulfites, Aromatics	2312-35-8	Flammable w/toxic products
Propargyl alcohol	Alcohols and glycols	107-19-7	Flammable Poison
Propionic acid	Acidic compounds, organic	79-09-4	Combustible
Propionic anhydride	Acidic compounds, organic	123-62-6	Combustible
n-Propylamine	Amines, alkyl	107-10-8	Flammable w/toxic products
Propylene dichloride	Aliphatics, halogenated	78-87-5	Flammable w/toxic products
Propylene oxide	Oxides, alkylene	75-56-9	Flammable Corrosive
1,2-Propylenimine	Amines, alkyl	75-55-8	Flammable w/toxic products Potential carcinogen
Pyridine	Amines, aryl	110-86-1	Flammable w/toxic products
Quinoline	Amines, aryl	91-22-5	Combustible
Safrole	Aromatics, Ethers	94-59-7	Combustible Potential carcinogen
Selenious acid	Acidic compounds, inorganic	7783-00-8	
Sodium hypochlorite (in aqueous solution)	Basic compounds	10022-70-5 7681-52-9	
Styrene	Aromatics, Olefins	100-42-5	Flammable Polymerizable
Sulfur monochloride	Halides, inorganic	12771-08-3	Corrosive Reactive
Sulfuric acid	Acidic compounds, inorganic	7664-93-9	Corrosive Reactive
2,4,5-T amines	Amines, aryl, Aromatics, halogenated	2008-46-0	Combustible w/toxic products

TABLE 2. LIQUIDS RELEASED ON LAND

Hazardous Substance	Chemical Class	CAS No.	Hazard(s), in Addition to Toxicity
2,4,5-T esters	Esters, Aromatics, halogenated	N.A.	
1,1,1,2-Tetrachloro-ethane	Aliphatics, halogenated	630-20-6	
1,1,2,2-Tetrachloro-ethane	Aliphatics, halogenated	79-34-5	
Tetrachloroethylene	Aliphatics, halogenated, Olefins	127-18-4	
Tetraethyldithiopyro-phosphate	Organophosphates	3689-24-5	Poison
Tetraethyl lead	Organometallics, Heavy metals	78-00-2	Combustible w/toxic products Poison
Tetraethyl pyrophosphate	Organophosphates	107-49-3	Combustible w/toxic products Poison
Tetrahydrofuran	Ethers	109-99-9	Flammable
Tetranitromethane	Nitro compounds	509-14-8	Explosive Oxidizer
Thiophenol	Phenols and cresols	108-98-5	
Toluene	Aromatics	108-88-3	Flammable
Toluene diisocyanate	Cyanates, Aromatics	584-84-9	Combustible w/toxic products Poison
2,4,5-TP acid esters	Esters, Aromatics, halogenated	N.A.	Combustible w/toxic products
1,2,4-Trichlorobenzene	Aromatics, halogenated	120-82-1	Combustible w/toxic products
1,1,2-Trichloroethane	Aliphatics, halogenated	79-00-5	Combustible w/toxic products
Trichloroethylene	Aliphatics, halogenated	79-01-6	(Flammable w/toxic products)

TABLE 2 . LIQUIDS RELEASED ON LAND

Hazardous Substance	Chemical Class	CAS No.	Hazard(s), in Addition to Toxicity
Trichloromonofluoro-methane	Aliphatics, halogenated	75-69-4	
Triethylamine	Amines, alkyl	121-44-8	Flammable w/toxic products Corrosive
Vinyl acetate	Esters, Olefins	108-05-4	Flammable Polymerizable
Vinylidene chloride	Aliphatics, halogenated	75-35-4	Flammable w/toxic products Polymerizable Potential carcinogen
Xylene	Aromatics	1330-20-7	Flammable
Xylenol	Phenols and cresols	1300-71-6	Combustible

Note 1: Specific gravity calculated according to Grains Method given in the Handbook of Chemical Property Estimation Methods.

Note 2: Specific gravity and/or soluility estimated based on active chemical groups and physical characteristics of structurally similar compounds.

LEGEND: N/A = Not available

TABLE 3. PARTICULATE SOLIDS RELEASED ON LAND

Hazardous Substance	Chemical Class	CAS No.	Hazard(s), in Addition to Toxicity
Acenaphthene	Aromatics	83-32-9	Combustible
Acenaphthylene	Aromatics	208-96-8	Combustible
2-Acetylaminofluorene	Amines, aryl	53-96-3	Potential carcinogen
1-Acetyl-2-thiourea	Ureas	591-08-2	
Acrylamide	Amides, anilides and imides	79-06-1	Polymerizable
Adipic acid	Acidic compounds, organic	124-04-9	
Aldicarb	Esters	116-06-3	
Aldrin	Aromatics, halogenated	309-00-2	Combustible w/toxic products Potential carcinogen Poison
Aluminum phosphide	Phosphorous and compounds	20859-73-8	Flammable w/toxic products Reactive
Aluminum sulfate	Sulfates	10043-01-3	
5-(Aminomethyl)-3-isoxazolol	Amine, alkyl	2763-96-4	
Amitrole	Azo compounds	61-82-5	Potential carcinogen
Ammonium acetate	Organic ammonium compounds	631-61-8	
Ammonium benzoate	Organic ammonium compounds	1863-63-4	Combustible w/toxic products
Ammonium bicarbonate	Organic ammonium compounds	1066-33-7	
Ammonium bichromate	Chromates	7789-09-5	Corrosive Oxidizer Flammable
Ammonium bifluoride	Halides, inorganic	1341-49-7	Corrosive
Ammonium bisulfite	Sulfites	10192-30-0	

TABLE 3. PARTICULATE SOLIDS RELEASED ON LAND

Hazardous Substance	Chemical Class	CAS No.	Hazard(s), in Addition to Toxicity
Ammonium carbamate	Organic ammonium compounds	1111-78-0	
Ammonium carbonate	Organic ammonium compounds	10361-29-2	
Ammonium chloride	Halides, inorganic	12125-02-9	
Ammonium chromate	Chromates	7788-98-9	
Ammonium citrate, dibasic	Organic ammonium compounds	3012-65-5	
Ammonium fluoborate	Organic ammonium compounds	13826-83-0	Corrosive
Ammonium fluoride	Halides, inorganic	12125-01-8	Corrosive
Ammonium oxalate	Organic ammonium compounds	6009-70-7 5972-73-6 14258-49-2	
Ammonium picrate	Nitro compounds	131-74-8	Flammable w/toxic products Explosive
Ammonium silicofluoride	Halides, inorganic	16919-19-0	Corrosive
Ammonium sulfamate	Sulfones, sulfoxides, and sulfonates	7773-06-0	
Ammonium sulfite	Sulfites	10196-04-0	Combustible w/toxic products
Ammonium tartrate	Organic ammonium compounds	3164-29-2 14307-43-8	
Ammonium thiocyanate	Cyanates	1762-95-4	Combustible w/toxic products
Ammonium thiosulfate	Sulfates	7783-18-8	
Ammonium vanadate	Heavy metals	7803-55-6	
Anthracene	Aromatics	120-12-7	Combustible
Antimony	Heavy metals	7440-36-0	Combustible w/toxic products
Antimony potassium tartrate	Organometallics Heavy metals	28300-74-5	

TABLE 3. PARTICULATE SOLIDS RELEASED ON LAND

Hazardous Substance	Chemical Class	CAS No.	Hazard(s), in Addition to Toxicity
Antimony tribromide	Halides, inorganic Heavy metals	7789-61-9	Corrosive Reactive
Antimony trichloride	Halides, inorganic Heavy metals	10025-91-9	Corrosive Reactive
Antimony trifluoride	Halides, inorganic Heavy metals	7783-56-4	Corrosive Reactive
Antimony trioxide	Oxides Heavy metals	1309-64-4	
Arsenic	Heavy metals	7440-38-2	Combustible w/toxic products Potential carcinogen Poison
Arsenic acid	Acidic compounds, inorganic, Heavy metals	1327-52-2 7778-39-4	Corrosive Poison
Arsenic disulfide	Sulfides and mercaptans Heavy metals	1303-32-8	Combustible w/toxic products Poison
Arsenic pentoxide	Oxides Heavy metals	1303-28-2	Corrosive Poison
Arsenic trioxide	Oxides Heavy metals	1327-53-3	Corrosive Poison
Arsenic trisulfide	Sulfides and mercaptans Heavy metals	1303-33-9	Combustible w/toxic products Poison
Asbestos	(See asbestos)	1332-21-4	Potential carcinogen
Auramine	Amines, aryl	492-80-8	Potential carcinogen
Azaserine	Azo compounds	115-02-6	Potential carcinogen
Barium cyanide	Cyanides and nitriles	542-62-1	Poison
3,4-Benzacridine	Aromatics	225-51-4	

TABLE 3. PARTICULATE SOLIDS RELEASED ON LAND

Hazardous Substance	Chemical Class	CAS No.	Hazard(s), in Addition to Toxicity
1,2-Benzanthracene	Aromatics	56-55-3	Potential carcinogen
alpha-Benzenehexa-chloride	Aliphatics, halogenated	319-84-6	Potential carcinogen
beta-Benzenehexa-chloride	Aliphatics, halogenated	319-85-7	Potential carcinogen
delta-Benzenehexa-chloride	Aliphatics, halogenated	319-86-8	Potential carcinogen
Benzidine	Amines, aryl	92-87-5	Combustible w/toxic products Potential carcinogen Poison
Benzo[b]fluoranthene	Aromatics	205-99-2	Potential carcinogen
Benzo[k]fluoranthene	Aromatics	207-08-9	Potential carcinogen
Benzo[ghi]perylene	Aromatics	191-24-2	
Benzoic acid	Acidic compounds, organic	65-85-0	Combustible
Benzo[a]pyrene	Aromatics	50-32-8	Potential carcinogen
p-Benzoquinone	Ketones	106-51-4	Combustible
Beryllium chloride	Halides, inorganic Heavy metals	7787-47-5	Poison
Beryllium	Heavy metals	7440-41-7	Flammable w/toxic products Potential carcinogen
Beryllium fluoride	Halides, inorganic Heavy metals	7787-49-7	Poison
Beryllium nitrate	Nitrates and nitrites Heavy metals	7787-55-5 13597-99-4	Oxidizer

TABLE 3. PARTICULATE SOLIDS RELEASED ON LAND

Hazardous Substance	Chemical Class	CAS No.	Hazard(s), in Addition to Toxicity
Brucine	(See strychnine and salts)	357-57-3	Combustible w/toxic products Poison
Cacodylic acid	Organometallics Heavy metals	75-60-5	Poison
Cadmium	Heavy metals	7440-43-9	Flammable w/toxic products Potential carcinogen
Cadmium acetate	Organometallics Heavy metals	543-90-8	Poison
Cadmium bromide	Halides, inorganic Heavy metals	7789-42-6	Poison
Cadmium chloride	Halides, inorganic Heavy metals	10108-64-2	Potential carcinogen
Calcium arsenate	Heavy metals	7778-44-1	Poison
Calcium arsenite	Heavy metals	52740-16-6	Poison
Calcium carbide	Organometallics	75-20-7	Flammable Reactive
Calcium chromate	Chromate	13765-19-0	Potential carcinogen
Calcium cyanide	Cyanides and nitriles	592-01-8	Reactive Poison
Calcium dodecylbenzene sulfonate	Sulfones, sulfoxides, and sulfonates	26264-06-2	
Calcium hypochlorite	Halides, inorganic	7778-54-3	Oxidizer
Captan	Acidic compounds, organic, Amides, anilides, and imides	133-06-2	Combustible w/toxic products
Carbaryl	Esters	63-25-2	Combustible w/toxic products
Carbofuran	Esters	1563-66-2	Combustible w/toxic products Poison

TABLE 3. PARTICULATE SOLIDS RELEASED ON LAND

Hazardous Substance	Chemical Class	CAS No.	Hazard(s), in Addition to Toxicity
Chlorambucil	Aromatics, halogenated Amines, aryl	305-03-3	Potential carcinogen
Chlordane	Aliphatics, halogenated Olefins	57-74-9	Combustible w/toxic products Potential carcinogen
Chlornaphazine (note1)	Amines, aryl	494-03-1	Potential carcinogen
p-Chloroaniline	Aromatics, halogenated Amines, aryl	106-47-8	
p-Chloro-m-cresol	Phenols and cresols Aromatics, halogenated	59-50-7	Combustible w/toxic products
2-Chloronaphthalene	Aromatics, halogenated	91-58-7	Combustible w/toxic products
1-(o-Chlorophenyl)-thiourea	Aromatics, halogenated Ureas	5344-82-1	
4-Chloro-o-toluidine, hydrochloride	Aromatics, halogenated Amines, aryl	3165-93-3	Poison
Chlorpyrifos	Organophosphates Aromatics, halogenated	2921-88-2	Combustible w/toxic products
Chromic acetate	Organometallics	1066-30-4	
Chromic acid	Acidic compounds, inorganic	11115-74-5	Corrosive Oxidizer Potential carcinogen
Chromic sulfate	Sulfates	10101-53-8	
Chromium	Heavy metals	7440-47-3	Flammable w/toxic products
Chromous chloride	Halides, inorganic	10049-05-5	Reactive
Chrysene	Aromatics	218-01-9	Combustible
Cobaltous bromide	Halides, inorganic Heavy metals	7789-43-7	
Cobaltous formate	Organometallics Heavy metals	544-18-3	

TABLE 3. PARTICULATE SOLIDS RELEASED ON LAND

Hazardous Substance	Chemical Class	CAS No.	Hazard(s), in Addition to Toxicity
Cobaltous sulfamate	Heavy metals	14017-41-5	
Copper	Heavy metals	7440-50-8	
Copper cyanide	Cyanides and nitriles Heavy metals	544-92-3	Poison
Coumaphos	Organophosphates Aromatics, halogenated	56-72-4	Combustible w/toxic products Poison
Cupric acetate	Organometallics Heavy metals	142-71-2	
Cupric acetoarsenite	Organometallics Heavy metals	12002-03-8	Poison
Cupric chloride	Halides, inorganic Heavy metals	7447-39-4	
Cupric nitrate	Nitrates and nitriles Heavy metals	3251-23-8	
Cupric oxalate	Organometallics Heavy metals	814-91-5	
Cupric sulfate	Sulfates Heavy metals	7758-98-7	
Cupric sulfate ammoniated	Sulfates Heavy metals	10380-29-7	
Cupric tartrate	Organometallics Heavy metals	815-82-7	
Cyanides (soluble salts and complexes)	Cyanides and nitriles	57-12-5	Poison
Cyanogen bromide	Cyanides and nitriles	506-68-3	Poison
Cyclophosphamide	Organophosphates Amides, anilides and imides	50-18-0	Potential carcinogen
2,4-D acid	Acidic compounds, organic, Aromatics, halogenated	94-75-7	Combustible w/toxic products
Daunonycin	Aromatics Ketones	20830-81-3	Potential carcinogen

TABLE 3. PARTICULATE SOLIDS RELEASED ON LAND

Hazardous Substance	Chemical Class	CAS No.	Hazard(s), in Addition to Toxicity
DDD	Aromatics, halogenated	72-54-8	Combustible w/toxic products Potential carcinogen
DDE	Aromatics, halogenated	72-55-9	Combustible w/toxic products Potential carcinogen
DDT	Aromatics, halogenated	50-29-3	Combustible w/toxic products Potential carcinogen
Diallate	Esters	2303-16-4	
Dibenzo[a,h]anthracene	Aromatics	53-70-3	Potential carcinogen
Dibenzo[a,i]pyrene	Aromatics	189-55-9	Potential carcinogen
Dicamba	Acidic compounds, organic, Aromatics, halogenated	1918-00-9	Combustible w/toxic products
Dichlobenil	Cyanides and nitriles Aromatics, halogenated	1194-65-6	Combustible w/toxic products
Dichlone	Aromatics, halogenated	117-80-6	Combustible w/toxic products
p-Dichlorobenzene	Aromatics, halogenated	106-46-7	Combustible w/toxic products
3,3'-Dichlorobenzidine	Aromatics, halogenated Amines, aryl	91-94-1	Combustible w/toxic products Potential carcinogen
2,4-Dichlorophenol	Phenols and cresols Aromatics, halogenated	120-83-2	Combustible w/toxic products
2,6-Dichlorophenol	Phenols and cresols Aromatics, halogenated	87-65-0	Combustible w/toxic products
Dieldrin	Epoxides Aromatics, halogenated	60-57-1	Potential carcinogen

TABLE 3. PARTICULATE SOLIDS RELEASED ON LAND

Hazardous Substance	Chemical Class	CAS No.	Hazard(s), in Addition to Toxicity
Diethylstilbestrol	Aromatics Ketones	56-53-1	Potential carcinogen
Dimethoate	Organophosphates	60-51-5	Flammable w/toxic products
3,3'-Dimethoxybenzidine	Amines, aryl	119-90-4	Potential carcinogen
Dimethylaminoazobenzene	Azo compounds Amines, aryl	60-11-7	Potential carcinogen
7,12-Dimethylbenz[a]-anthracene	Aromatics	57-97-6	
3,3'-Dimethylbenzidine	Amines, aryl	119-93-7	Potential carcinogen
2,4-Dimethylphenol	Phenols and cresols	105-67-9	Combustible
Dinitrobenzene (mixed)	Nitro compounds Aromatics	25154-54-5	Poison
4,6-Dinitro-o-cresol	Nitro compounds Phenols and cresols	534-52-1	Combustible w/toxic products
4,6-Dinitro-o-cyclo-hexylphenol	Nitro compounds Phenols and cresols	131-89-5	Combustible w/toxic products
Dinitrophenol	Nitro compounds Phenols and cresols	25550-58-7	Combustible w/toxic products Poison
2,4-Dinitrophenol	Nitro compounds Phenols and cresols	51-28-5	Combustible w/toxic products Poison
Dinitrotoluene	Nitro compounds Aromatics	25321-14-6	
2,4-Dinitrotoluene	Nitro compounds Aromatics	121-14-2	
2,6-Dinitrotoluene	Nitro compounds Aromatics	606-20-2	
Dinoseb	Nitro compounds Phenols and cresols	88-85-7	Flammable w/toxic products

TABLE 3. PARTICULATE SOLIDS RELEASED ON LAND

Hazardous Substance	Chemical Class	CAS No.	Hazard(s), in Addition to Toxicity
1,2-Diphenylhydrazine	Hydrazines and Hydrazides	122-66-7	Combustible w/toxic products Potential carcinogen
Diquat	Aromatics, halogenated	85-00-7 2764-72-9	Combustible w/toxic products
Disulfoton	Organophosphates	298-04-4	Combustible w/toxic products Poison
2,4-Dithiobiuret	Amides, anilides, and imides	541-53-7	
Diuron	Ureas Aromatics, halogenated	330-54-1	Combustible w/toxic products
Endosulfan	Aromatics, halogenated Sulfones, sulfoxides, and sulfonates	115-29-7	Combustible w/toxic products Poison
alpha-Endosulfan	Aromatics, halogenated Sulfones, sulfoxides, and sulfonates	959-98-8	Poison
beta-Endosulfan	Aromatics, halogenated Sulfones, sulfoxides and sulfonates	33213-65-9	Poison
Endosulfan sulfate	Aromatics, halogenated Sulfones, sulfoxides and sulfonates	1031-07-8	Combustible w/toxic products
Endothall	Acidic compounds, organic	145-73-3	
Endrin	Epoxides Aromatics, halogenated	72-20-8	Poison
Epinephrene	Amines, aryl	51-43-4	
Ethyl carbamate	Esters	51-79-6	Potential carcinogen
Ethylenebis(dithio-carbamic acid)	Acidic compounds, organic	111-54-6	

TABLE 3. PARTICULATE SOLIDS RELEASED ON LAND

Hazardous Substance	Chemical Class	CAS No.	Hazard(s), in Addition to Toxicity
Ethylenediamine tetra-acetic acid	Amines, alkyl Acidic compounds, organic	60-00-4	Combustible w/toxic products
Ethylenethiourea	Ureas	96-45-7	Potential carcinogen
Famphur	Organophosphates Amides, anilides and imides	52-85-7	
Ferric ammonium citrate	Organometallics	1185-57-5	
Ferric ammonium oxalate	Organometallics	2944-67-4 55488-87-4	
Ferric chloride	Halides, inorganic	7705-08-0	
Ferric fluoride	Halides, inorganic	7783-50-8	
Ferric nitrate	Nitrates and nitrites	10421-48-4	Oxidizer
Ferric sulfate	Sulfates	10028-22-5	
Ferrous ammonium sulfate	Sulfates	10045-89-3	
Ferrous chloride	Halides, inorganic	7758-94-3	
Ferrous sulfate	Sulfates	7720-78-7 7782-63-0	
Fluoranthene	Aromatics	206-44-0	
Fluorene	Aromatics	86-73-7	
Fluoroacetamide	Amides, anilides, and imides	640-19-7	
Fumaric acid	Acidic compounds, organic	110-17-8	Combustible
Guthion	Aromatics Organophosphates	86-50-0	Poison
Heptachlor	Aliphatics, halogenated Olefins	76-44-8	Potential carcinogen
Heptachlor epoxide	Aliphatics, halogenated Epoxides	1024-57-3	Potential carcinogen

TABLE 3. PARTICULATE SOLIDS RELEASED ON LAND

Hazardous Substance	Chemical Class	CAS No.	Hazard(s), in Addition to Toxicity
Hexachlorobenzene	Aromatics, halogenated	118-74-1	Potential carcinogen
Hexachloroethane	Aliphatics, halogenated	67-72-1	
Hexachlorohexahydro-endo,endo-dimethano-naphthalene	Aromatics, halogenated	465-73-6	Poison
Hexachlorophene	Aromatics, halogenated	70-30-4	
Indeno(1,2,3-cd)pyrene	Aromatics	193-39-5	Potential carcinogen
Iron dextran	Organometallics	9004-66-4	Potential carcinogen
Isopropanolamine dodec-ylbenzenesulfonate	Sulfones, sulfoxides and sulfonates	42504-46-1	Combustible w/toxic products
Kelthane	Aromatics, halogenated	115-32-2	Combustible w/toxic products
Kepone	Aliphatics, halogenated Ketones	143-50-0	Combustible w/toxic products Potential carcinogen
Lasiocarpine	Acidic compounds, organic	303-34-4	Potential carcinogen
Lead	Heavy metals	7439-92-1	
Lead acetate	Organometallics Heavy metals	301-04-2	Potential carcinogen
Lead arsenate	Heavy metals	7784-40-9 3687-31-8 7645-25-2 10102-48-4	Poison
Lead chloride	Halides, inorganic Heavy metals	7758-95-4	
Lead fluoride	Halides, inorganic Heavy metals	7783-46-2	
Lead iodide	Halides, inorganic Heavy metals	10101-63-0	

TABLE 3. PARTICULATE SOLIDS RELEASED ON LAND

Hazardous Substance	Chemical Class	CAS No.	Hazard(s), in Addition to Toxicity
Lead nitrate	Nitrates and nitrites Heavy metals	10099-74-8	Oxidizer
Lead phosphate	Phosphates and phosphonates Heavy metals	7446-27-7	Potential carcinogen
Lead stearate	Organometallics Heavy metals	7428-48-0 1072-35-1 56189-09-4	
Lead subacetate	Organometallics Heavy metals	1335-32-6	Potential Carcinogen
Lead sulfate	Sulfates Heavy metals	7446-14-2 15739-80-7	
Lead sulfide	Sulfides and mercaptans Heavy metals	1314-87-0	
Lead thiocyanate	Cyanates Heavy metals	592-87-0	
Lindane	Aliphatics, halogenated	58-89-9	Potential carcinogen
Lithium chromate	Chromates	14307-35-8	
Maleic acid	Acidic compounds, organic	110-16-7	
Maleic anhydride	Acidic compounds, organic	108-31-6	Combustible
Maleic hydrazide	Hydrazines and hydrazides	123-33-1	Combustible w/toxic products
Malononitrile	Cyanides and nitriles	109-77-3	Combustible w/toxic products
Melphalan	Aromatics, halogenated Amines, aryl	148-82-3	Potential carcinogen
Mercaptodimethur	Sulfides and mercaptans	2032-65-7	
Mercuric cyanide	Cyanides and nitriles Heavy metals	592-04-1	Poison
Mercuric nitrate	Nitrates and nitrites Heavy metals	10045-94-0	Oxidizer

TABLE 3. PARTICULATE SOLIDS RELEASED ON LAND

Hazardous Substance	Chemical Class	CAS No.	Hazard(s), in Addition to Toxicity
Mercuric sulfate	Sulfates Heavy metals	7783-35-9	Poison
Mercuric thiocyanate	Cyanates Heavy metals	592-85-8	Poison
Mercurous nitrate	Nitrates and nitrites Heavy metals	7782-86-7 10415-75-5	Oxidizer
Mercury fulminate	Cyanides and nitriles	628-86-4	Explosive
Methomyl	Amides, anilides and imides	16752-77-5	
Methyl bromide	Halides, alkyl	74-83-9	Combustible w/toxic products Poison
Methoxychlor	Aromatics, halogenated	72-43-5	
3-Methylcholanthrene	Aromatics	56-49-5	Potential carcinogen
4,4'-Methylenebis-(2-chloroaniline)	Aromatics, halogenated Amines, aryl	101-14-4	Potential carcinogen
Methyl parathion	Organophosphates	298-00-0	Combustible w/toxic products Poison
N-Methyl-N'-nitro-N-nitrosoguanidine	Nitro compounds Nitroso compounds	70-25-7	Flammable w/toxic products Potential carcinogen
Methylthiouracil	Amines, akyl	56-04-2	Potential carcinogen
Mexacarbate	Esters	315-18-4	Combustible w/toxic products Poison
Mitomycin C	(See mitomycin)	50-07-7	Potential carcinogen
Naled	Aliphatics, halogenated Organophosphates	300-76-5	Combustible w/toxic products
Naphthalene	Aromatics	91-20-3	Combustible

TABLE 3. PARTICULATE SOLIDS RELEASED ON LAND

Hazardous Substance	Chemical Class	CAS No.	Hazard(s), in Addition to Toxicity
1,4-Naphthoquinone	Aromatics	130-15-4	
1-Naphthylamine	Amines, aryl	134-32-7	
2-Naphthylamine	Amines, aryl	91-59-8	Potential carcinogen
alpha-Naphthylthiourea	Ureas	86-88-4	
Nickel	Heavy metals	7440-02-0	Flammable w/toxic products Potential carcinogen
Nickel ammonium sulfate	Sulfates Heavy metals	15699-18-0	
Nickel chloride	Halides, inorganic Heavy metals	37211-05-5 7718-54-9	
Nickel cyanide	Cyanides and nitriles Heavy metals	557-19-7	Poison
Nickel hydroxide	Basic compounds Heavy metals	12054-48-7	
Nickel nitrate	Nitrates and nitrites Heavy metals	14216-75-2	Oxidizer
Nickel sulfate	Sulfates Heavy metals	7786-81-4	
Nicotine salts	Amines, aryl	54-11-5	Combustible w/toxic products Poison
p-Nitroaniline	Nitro compounds Amines, aryl	100-01-6	Poison
Nitroglycerine	Nitro compounds	55-63-0	Flammable w/toxic products Explosive
Nitrophenol (mixed)	Nitro compounds Phenols and cresols	25154-55-6	Combustible w/toxic products
2-Nitrophenol	Nitro compounds Phenols and cresols	88-75-5	Combustible w/toxic products

TABLE 3. PARTICULATE SOLIDS RELEASED ON LAND

Hazardous Substance	Chemical Class	CAS No.	Hazard(s), in Addition to Toxicity
4-Nitrophenol	Nitro compounds Phenols and cresols	100-02-7	Combustible w/toxic products
N-Nitrosodi-n-butylamine	Nitroso compounds	924-16-3	Combustible w/toxic products Potential carcinogen
N-Nitrosodiphenylamine	Nitroso compounds Amine, aryl	86-30-6	Potential carcinogen
N-Nitroso-N-ethylurea	Nitroso compounds	759-73-9	Potential carcinogen
N-Nitroso-N-methyl-urethane	Nitroso compounds	615-53-2	Combustible w/toxic products Potential carcinogen
N-Nitrosopiperidine	Nitroso compounds	100-75-4	Potential carcinogen
Nitrotoluene	Nitro compounds Aromatics	1321-12-6	Combustible w/toxic products
5-Nitro-o-toluidine	Nitro compounds Amines, aryl	99-55-8	
Osmium tetroxide	Oxides Heavy metals	20816-12-0	
Paraformaldehyde	Aldehydes	30525-89-4	Combustible
Pentachlorobenzene	Aromatics, halogenated	608-93-5	Combustible w/toxic products
Pentachloronitrobenzene	Nitro compounds Aromatics, halogenated	82-68-8	
Pentachlorophenol	Phenols and cresols Aromatics, halogenated	87-86-5	
Phenacetin	Aromatics Amides, anilides and imides	62-44-2	Potential carcinogen
Phenanthrene	Aromatics	85-01-8	

TABLE 3. PARTICULATE SOLIDS RELEASED ON LAND

Hazardous Substance	Chemical Class	CAS No.	Hazard(s), in Addition to Toxicity
Phenol	Phenols and cresols	108-95-2	Combustible Corrosive Poison
Phenyl dichloroarsine	Aromatics, halogenated Heavy metals	696-28-6	Poison
Phenylmercuric acetate	Organometallics Heavy metals	62-38-4	Combustible w/toxic products
N-Phenylthiourea	Ureas Aromatics	103-85-5	
Phorate	Organophosphates	298-02-2	
Phosphorus	Phosphorous and compounds	7723-14-0	Flammable w/toxic products Poison
Phosphorus pentasulfide	Phosphorous and compounds Sulfides and mercaptans	1314-80-3	Flammable w/toxic products Reactive
Phthalic anhydride	Aromatics	85-44-9	Corrosive
Potassium arsenate	Heavy metals	7784-41-0	Poison
Potassium arsenite	Heavy metals	10124-50-2	Poison
Potassium bichromate	Chromates	7778-50-9	Corrosive Oxidizer
Potassium chromate	Chromates	7789-00-6	
Potassium cyanide	Cyanides and nitriles	151-50-8	Poison
Potassium hydroxide	Basic compounds	1310-58-3	Corrosive
Potassium permanganate	Basic compounds	7722-64-7	Corrosive Oxidizer
Potassium silver cyanide	Cyanides and nitriles Heavy metals	506-61-6	
Pronamide	Aromatics, halogenated Amides,anilides, and imides	23950-58-5	
1,3-Propane sultone	Sulfones, sulfoxides, and sulfonates	1120-71-4	Potential carcinogen

TABLE 3. PARTICULATE SOLIDS RELEASED ON LAND

Hazardous Substance	Chemical Class	CAS No.	Hazard(s), in Addition to Toxicity
Pyrene	Aromatics	129-00-0	Combustible w/toxic products
Pyrethrins	Acidic compounds, organic	121-21-1 121-29-9	Combustible w/toxic products
4-Pyridinamine	Amines, aryl	504-24-5	Combustible w/toxic products
Reserpine	Aromatics	50-55-5	
Resorcinol	Aromatics	108-46-3	Combustible
Saccharin	Aromatics Sulfones, sulfoxides and sulfonates	81-07-2	Potential carcinogen
Selenium	Heavy metals	7782-49-2	Combustible w/toxic products
Selenium disulfide	Sulfides and mercaptans Heavy metals	7488-56-4	Reactive
Selenium oxide	Oxides Heavy metals	7446-08-4	Poison
Selenourea	Ureas Heavy metals	630-10-4	
Silver	Heavy metals	7440-22-4	
Silver cyanide	Cyanides and nitriles Heavy metals	506-64-9	Poison
Silver nitrate	Nitrates and nitrites Heavy metals	7761-88-8	Oxidizer
Sodium	Alkali metals	7440-23-5	Flammable Reactive Corrosive
Sodium arsenate	Heavy metals	7631-89-2	Poison
Sodium arsenite	Heavy metals	7784-46-5	Poison
Sodium azide	Azo compounds	26628-22-8	Explosive Combustible w/toxic products Poison

TABLE 3. PARTICULATE SOLIDS RELEASED ON LAND

Hazardous Substance	Chemical Class	CAS No.	Hazard(s), in Addition to Toxicity
Sodium bichromate	Chromates	10588-01-9	Corrosive Oxidizer
Sodium bifluoride	Halides, inorganic	1333-83-1	Corrosive
Sodium bisulfite	Sulfites	7631-90-5	
Sodium chromate	Chromates	7775-11-3	
Sodium cyanide	Cyanides and nitriles	143-33-9	Poison
Sodium dodecylbenzene sulfonate	Sulfones, sulfoxides and sulfonates	25155-30-0	
Sodium fluoride	Halides, inorganic	7681-49-4	
Sodium fluoroacetate	Organometallics	62-74-8	
Sodium hydrosulfide	Sulfides and mercaptans	16721-80-5	Flammable w/toxic products Reactive
Sodium hydroxide	Basic compounds	1310-73-2	Corrosive Reactive
Sodium methylate	Organometallics	124-41-4	Flammable w/toxic products Reactive
Sodium nitrite	Nitrates and nitrites	7632-00-0	Oxidizer
Sodium phosphate, dibasic	Phosphates and phosphonates	7558-79-4 10028-24-7 10039-32-4 10140-65-5	
Sodium phosphate, tribasic	Phosphates and phosphonates	7601-54-9 7785-84-4 10101-89-0 10361-89-4 7758-29-4 10124-56-8	
Sodium selenite	Heavy metals	10102-18-8 7782-82-3	Poison
Streptozotocin	(See streptozotocin)	18883-66-4	Potential carcinogen

TABLE 3. PARTICULATE SOLIDS RELEASED ON LAND

Hazardous Substance	Chemical Class	CAS No.	Hazard(s), in Addition to Toxicity
Strontium chromate	Chromates Heavy metals	7789-06-2	
Strontium sulfide	Sulfides and mercaptans Heavy metals	1314-96-1	Reactive
Strychnine and salts	(See strychnine and salts)	57-24-9	Poison
2,4,5-T acid	Aromatics, halogenated	93-76-5	
2,4,5-T salts	Aromatics, halogenated	13560-99-1	Combustible w/toxic products
1,2,4,5-Tetrachloro-benzene	Aromatics, halogenated	95-94-3	
2,3,7,8-tetrachloro-dibenzo-p-dioxin	Aromatics, halogenated	1746-01-6	Combustible w/toxic products Potential carcinogen
2,3,4,6-Tetrachloro-phenol	Phenols and cresols Aromatics, halogenated	58-90-2	Combustible w/toxic products
Thallium	Heavy metals	7440-28-0	
Thallium(I) acetate	Organometallics Heavy metals	563-68-8	
Thallium(I) carbonate	Organometallics Heavy metals	6533-73-9	
Thallium(I) chloride	Halides, inorganic Heavy metals	7791-12-0	Poison
Thallium(I) nitrate	Nitrates and nitrites Heavy metals	10102-45-1	Oxidizer
Thallium(III) oxide	Oxides Heavy metals	1314-32-5	Oxidizer
Thallium(I) selenide	Heavy metals	12039-52-0	
Thallium(I) sulfate	Sulfates Heavy metals	7446-18-6	Poison
Thioacetamide	Amides, anilides,and imides	62-55-5	Potential carcinogen

TABLE 3. PARTICULATE SOLIDS RELEASED ON LAND

Hazardous Substance	Chemical Class	CAS No.	Hazard(s), in Addition to Toxicity
Thiofanox	Sulfides and mercaptans Amides, anilides, and imides	39196-18-4	
Thiosemicarbazide	Azo compounds Ureas	79-19-6	
Thiourea	Ureas	62-56-6	Potential carcinogen
Thiram	Sulfides and mercaptans	137-26-8	Combustible w/toxic products
Toluenediamine	Amines, aryl	95-80-7	Potential carcinogen
o-Toluidine hydro-chloride	Aromatics, halogenated Amines, aryl	636-21-5	Potential carcinogen
Toxaphene	Aliphatics, halogenated	8001-35-2	Combustible w/toxic products Potential carcinogen
2,4,5-TP acid	Acidic compounds, organic Aromatics, halogenated	93-72-1	
Trichlorfon	Aliphatics, halogenated Organophosphates	52-68-6	
Trichloromethane-sulfenyl chloride	Aliphatics, halogenated	594-42-3	Poison
Trichlorophenol	Phenols and cresols Aromatics, halogenated	25167-82-2	
2,4,5-Trichlorophenol	Phenols and cresols Aromatics, halogenated	95-95-4	Combustible w/toxic products
2,4,6-Trichlorophenol	Phenols and cresols Aromatics, halogenated	88-06-2	Combustible w/toxic products
Triethanolamine dodecyl-benzene sulfonate	Sulfones, sulfoxides, and sulfonates	27323-41-7	
sym-Trinitrobenzene	Nitro compounds Aromatics	99-35-4	Flammable w/toxic products Explosive

TABLE 3. PARTICULATE SOLIDS RELEASED ON LAND

Hazardous Substance	Chemical Class	CAS No.	Hazard(s), in Addition to Toxicity
Tris(2,3-dibromopropyl) phosphate	Phosphates and phosphonates	126-72-7	Potential carcinogen
Trypan blue	Azo compounds	72-57-1	Potential carcinogen
Uracil mustard	Aliphatics, halogenated Amines, alkyl	66-75-1	Potential carcinogen
Uranyl acetate	Organometallics Heavy metals	541-09-3	Radioactive
Uranyl nitrate	Nitrates and nitrites Heavy metals	10102-06-4 36478-76-9	Radioactive Oxidizer
Vanadium pentoxide	Oxides Heavy metals	1314-62-1	
Vanadyl sulfate	Sulfates Heavy metals	27774-13-6	
Warfarin	Aromatics	81-81-2	
Zinc	Heavy metals	7440-66-6	Combustible
Zinc acetate	Organometallics Heavy metals	557-34-6	
Zinc ammonium chloride	Halides, inorganic Heavy metals	52628-25-8 14639-97-5 14639-98-6	
Zinc borate	Heavy metals	1332-07-6	
Zinc bromide	Halides, inorganic Heavy metals	7699-45-8	
Zinc carbonate	Organometallics Heavy metals	3486-35-9	
Zinc chloride	Halides, inorganic Heavy metals	7646-85-7	
Zinc cyanide	Cyanides and nitriles Heavy metals	557-21-1	Poison
Zinc fluoride	Halides, inorganic Heavy metals	7783-49-5	

TABLE 3. PARTICULATE SOLIDS RELEASED ON LAND

Hazardous Substance	Chemical Class	CAS No.	Hazard(s), in Addition to Toxicity
Zinc formate	Organometallics Heavy metals	557-41-5	
Zinc hydrosulfite	Sulfites Heavy metals	7779-86-4	
Zinc nitrate	Nitrates and nitrites Heavy metals	7779-88-6	Oxidizer
Zinc phenolsulfonate	Phenols and cresols Heavy metals	127-82-2	Combustible w/toxic products
Zinc phosphide	Phosphorous and compounds Heavy metals	1314-84-7	Flammable w/toxic products Reactive Poison
Zinc silicofluoride	Halides, inorganic Heavy metals	16871-71-9	
Zinc sulfate	Sulfates and sulfites Heavy metals	7733-02-0	
Zirconium nitrate	Nitrates and nitrites Heavy metals	13746-89-9	Oxidizer
Zirconium potassium fluoride	Halides, inorganic Heavy metals	16923-95-8	
Zirconium sulfate	Sulfates Heavy metals	14644-61-2	
Zirconium tetrachloride	Halides, inorganic Heavy metals	10026-11-6	Corrosive Reactive

TABLE 4. COMPRESSED GASES RELEASED INTO AIR

Hazardous Substance	Chemical Class	CAS No.	Hazard(s), in Addition to Toxicity
Ammonia	Ammonia	7664-41-7	Corrosive
Carbon oxyfluoride	Halides, alkyl	353-50-4	Reactive
Chlorine	Halogens	7782-50-5	Oxidizer Poison
Cyanogen	Cyanides and nitriles	460-19-5	Flammable w/toxic products Poison
Dichlorodifluoromethane	Halides, alkyl	75-71-8	
Dimethylamine	Amines, alkyl	124-40-3	Flammable w/toxic products Corrosive
Fluorine	Halogens	7782-41-4	Corrosive Reactive Oxidizer Poison
Formaldehyde	Aldehydes	50-00-0	Flammable
Hydrogen sulfide	Sulfides and mercaptans	7783-06-4	Flammable w/toxic products Poison
Methyl bromide	Halides,alkyl	74-83-9	Combustible w/toxic products Poison
Methyl chloride	Halides, alkyl	74-87-3	Flammable w/toxic products
Methyl mercaptan	Sulfides and mercaptans	74-93-1	Flammable w/toxic products Corrosive
Monomethylamine	Amines, alkyl	74-89-5	Flammable w/toxic products Corrosive
Nitric oxide	Oxides	10102-43-9	Poison Oxidizer
Phosgene	Halides, organic	75-44-5	Combustible w/toxic products Poison

TABLE 4. COMPRESSED GASES RELEASED INTO AIR

Hazardous Substance	Chemical Class	CAS No.	Hazard(s), in Addition to Toxicity
Phosphine	Phosphorous and compounds	7803-51-2	Flammable w/toxic products Poison
Trimethylamine	Amines, alkyl	75-50-3	Flammable w/toxic products Corrosive
Vinyl chloride	Halides, alkyl	75-01-4	Flammable w/toxic products Polymerizable Potential carcinogen

4. Fixed Facilities

This manual is directed principally toward smaller chemical manufacturing facilities; however, it can be applied to any fixed facility. If one excludes the petroleum refining industry, which is covered by an existing Oil Pollution Prevention Regulation (40 CFR 112), there are 20 major industry categories defined in 40 CFR 124, Appendix D, as follows:

No.	Major Industry Category
1	Timber Products Processing
2	Steam Electric Power Plants
3	Leather and Leather Products
4	Iron and Steel Manufacturing
5	Inorganic Chemicals Manufacturing
6	Textile Mills
7	Organic Chemicals Manufacturing
8	Nonferrous Metals Manufacturing
9	Paving and Roofing Materials
10	Paint and Ink Formulation and Printing
11	Soap and Detergent Manufacturing
12	Auto Wash and Other Laundries
13	Plastics and Synthetic Metals Manufacturing
14	Pulp, Paper and Board Mills, and Products Manufacturing
15	Rubber Processing
16	Miscellaneous Chemicals
17	Machinery and Mechanical Products Manufacturing
18	Electroplating
19	Ore Mining and Dressing
20	Coal Mining

Several of these categories fall within the chemical manufacturing area. Others, such as mining (Nos. 19 and 20), metal processing in various forms (Nos. 4, 8, 18, and 19), and production of non-metallics (Nos. 3, 6, 9, 13, 14, and 15), also deal with hazardous substances. Each industry has its own specialized equipment and facility areas. For the present purpose, this section lists and describes the equipment found in chemical manufacturing plants. For other industries, this section and the Appendix would have to be modified.

In accordance with Step 5 of the procedures for preparing a master plan, a list of plant component interactions with hazardous substances is required.

A check list of all facility areas and components is useful for developing the final list. Reference to the Chemical Engineers' Handbook (1973, 5th edition) led to the identification of major facility area categories which cover processing, transportation, storage, waste treatment, and disposal. A detailed description of hazardous substance interactive systems and equipment (including further division of the major categories) is provided in the Appendix. A listing of the systems and equipment is provided here:

1. Transport and storage of fluids (pumps, pipes, valves, tanks, etc.)
2. Handling of bulk and packaged solids (conveyors, silos, etc.)
3. Size reduction and enlargement (mills, compactors, etc.)
4. Heat generation and transport (fired process equipment, incinerators, etc.)
5. Heat transfer equipment (heat exchangers, condensers, etc.)
6. Evaporative cooling and refrigeration (cooling towers, cryogens, etc.)
7. Distillation columns
8. Gas absorption towers
9. Liquid extraction systems
10. Adsorption and ion exchange equipment
11. Miscellaneous separation processes (crystallization, membranes, etc.)
12. Liquid-gas systems (contacting, phase dispersion, phase separation)
13. Liquid-solid systems (contacting, phase dispersion, phase separation)
14. Gas-solid systems (contacting, phase dispersion, phase separation)
15. Liquid-liquid systems (contacting, phase dispersion, phase separation)
16. Solid-solid systems (contacting, phase dispersion, phase separation)
17. Waste treatment plants (equipment associated with primary, secondary, and tertiary treatment and disposal).

5. Facility Spill Prevention Practices

In general, spill prevention practices (SPPs) are independent of the exact nature of the facility, its processes and products. While most SPPs are based on common sense and experience and provide few new revelations, it is useful to have this compilation of procedures when preparing a spill prevention master plan.

SPPs have been drawn from government and industry reports and publications. Some, listed here, overlap into related areas such as response to a spill, maintenance, and training. If the facility already has plans and procedures in these areas, they can be adapted to the present purpose. A short discussion for each SPP listed follows.

SPILL PREVENTION ORGANIZATION

The facility spill prevention committee is responsible for developing the spill prevention master plan and assisting management in carrying out, maintaining, and improving the plan. The size of the committee should depend on the plant size and complexity of the situation. The committee should be headed by a part-time or full-time manager. For a very small facility handling few chemicals, part-time assistance from a larger plant may suffice. The initial task of the committee is to prepare a policy statement for management approval indicating overall goals and outlining specific activities such as:

- identifying all toxic and hazardous substances at the facility
- identifying all potential spill sources
- establishing incident reporting procedures
- assessing risks at different plant locations
- developing inspection and record procedures
- reviewing historical incidents and remedies
- coordinating incident response, cleanup, and notification procedures
- establishing a training program for plant and contractor personnel

- o assisting in interdepartmental spill prevention coordination
- o reviewing new construction and/or changes in processes and procedures

RISK IDENTIFICATION AND ASSESSMENT

All facility areas should be examined for potential risk of a hazardous substance release. In defining facility boundaries, a plant layout should be prepared. The areas should be drawn or mapped and a flow diagram of major process operations (Section 4) developed to indicate direction and quantity (volume) of materials flowing from one area to another. Storage areas and transportation terminals also should be located on the diagram and the direction of flow should be estimated. Topography is a significant factor in planning for prevention and safety. Liquids and heavier-than-air gases will flow downhill. Direction of prevailing winds should also be noted on the layout. Prevailing wind information will aid in keeping emitted flammable gases and vapors from potential ignition sources. Distance between the facility and facility boundary can enhance public safety and should be kept in mind during plant location. High-pressure vessels should be located as far from facility boundaries and public receptors as possible. Pressurized vessels and other hazardous materials should be isolated.

Detail the area map by identifying materials, quantities, location, and preventive systems (dikes, berms, dams, etc.). Sources of water for fire fighting, and locations of operating units, service and emergency (if any) facilities should be shown. Safety showers, hose cart stations, fire hydrants and extinguishers should all be located on one map. (These will all require periodic inspection.) Major outside community receptors should also be depicted on a more enlarged diagram. In a larger facility, a detailed three-dimensional scaled model showing facility operations and boundaries, would be appropriate to produce speedy and effective response. This is not necessary at smaller facilities.

A hazardous substance inventory should be taken (Section 3). Inflow, generation, storage, and outflow of all substances within the facility boundaries should be defined. This includes raw materials, products, byproducts, wastes, fuels, lubes, pesticides, disinfectants, etc. These should be incorporated into the facility mapping or on flow diagrams. Normal operating loads and overload conditions should also be included.

In risk assessment, the quantity of material, its degree of hazard (including whether single or multiple releases could occur), and its distance from the plant boundary and/or sensitive environment (e.g., a watercourse) are combined in a risk "score." Thus, high-risk areas can be located readily and given special attention. Physical, chemical, toxicological, and health data should be included in the investigation. Also, any possible synergisms or incompatibilities between materials stored, used, or transported at the facility should be considered. This includes cleaning and operation materials as well as chemicals used in processes. Another factor is the interrelation of various processes and activities that are carried out close to each other in space or time. For instance, use of solvents to clean equipment near electrical motors will be a high-risk operation if the motors are running and generating heat and sparks, but low-risk if the motors are shut down during cleanup.

Planned changes in materials inventory and processes should be examined beforehand as to possible risks, and design changes to eliminate risks should be adopted before activating new systems.

MATERIALS COMPATIBILITY

The specific requirement for materials compatibility should provide procedures to address the design and operation of equipment used in handling hazardous substances (Section 3, Appendix). Material compatibility encompasses three aspects: compatibility of the chemicals handled with container construction materials, compatibility of the mixture of chemicals in a container, and compatibility of the container with its environment.

The engineering practices in the plant should be evaluated and summarized with regard to corrosion and other aspects of material compatibility. Activities or processes that can possibly increase the risk potential of a substance should be evaluated relative to the materials involved. Practices such as mixing of chemicals and possible contact with various materials that may result in fire, explosion, or unusual corrosion or reactivity; cleaning of storage vessels and tanks; and equipment and materials testing should be evaluated and their potential for a release estimated. Recommendations for improved procedures should be made. Selected activities or processes to be considered and addressed include:

- Area washdowns
- Fuel transfers
- Heavy equipment operations
- Chemical transfers
- Storage tank fillings
- Cleaning operations
- Pretreatment processes
- Chemical sampling operations
- Treatment operations
- Pollutant removals
- Gas buildups in storage tanks
- Alterations of stored chemicals
- Heavy equipment operations

REPORTING AND RECORDKEEPING

An internal incident reporting system should be used to keep records of incidents such as controlled spills, leaks, runoff, etc., for the purpose of minimizing reoccurrence, expediting mitigation or cleanup activities, and complying with legal requirements. The primary reporting procedures should include notification of a discharge to appropriate plant personnel to initiate immediate action, and formal written reports for review and evaluation by management of the spill incident. Governmental and environmental agencies should be notified in the event that the spill or other reportable discharge reaches beyond facility boundaries. The incident reporting system should therefore, include the following elements:

- o Procedures for notifying appropriate plant personnel
  - Communications system for reporting incidents in-plant (e.g., telephone, alarms, radio, etc.)
  - Provisions to maintain communications in event of power failure
- o Action plan to prevent or mitigate incident
- o Recording of incidents for internal management review
- o Governmental and environmental agency notification of incident as required by law
- o Responsible party, company, and government officials
- o Names and telephone numbers of key personnel

Proper distribution and filing of these reports is important to the activities of the spill prevention committee, as well as the safe operation of the facility and the good health of its employees. The format for the report according to CERCLA 103(a) should be developed prior to incidents and be included in the plant's contingency plans. This information should be updated and available for review by EPA or state personnel. The format should include, but not be limited to:

- o Date and time of release
- o Type of release (solid, liquid, gas)
- o Location of release (be specific and use prepared plot diagrams)
- o Name(s) of substance(s) released
- o Quantity released (lbs) and reportable quantity information
- o Time and duration of release (hrs)

- Environmental media released to
  - air
  - surface water
  - groundwater
  - land surface
  - controlled/contained in impoundment
- Description of incident (known or suspected)
- Corrective measures taken (with times)
- Use of protective gear (if any)
- Whether release was reported to appropriate authorities
- Verification of notification (with explanations)

GOOD HOUSEKEEPING

Good housekeeping is the maintenance of a clean, orderly work environment that contributes to overall facility pollution control efforts. Occupational Safety and Health Administration (OSHA) includes housekeeping regulations in 29 CFR 1910, Sections 22(a), 141, and 176(c) that apply to industry, in general, and not specifically for toxic substances control. The principal elements in good housekeeping include proper storage of chemicals, prompt removal of spillage, floor maintenance, and unobstructed pathways and walkways. Methods for maintaining good housekeeping goals and employee interest in such are also key aspects.

Good housekeeping examples include: covering solid material storage piles and drum lots to prevent leaching by rainwater, and loading, storing, and unloading metal and fiber containers in a manner that will minimize the possibility of container damage. Chemicals of compatible nature must be stored in a neat, orderly, and easily accessible fashion. Chemicals stored in drums or bags should be staged according to chemical type. Storage vessels, tanks, sample containers, bags, pipes, barrels, etc., should be checked for integrity as well as compatibility with materials stored. Chemical staging must be in distinct areas, and the areas and types of chemicals stored must be labeled. An inventory and plot diagram of stored chemicals is needed to help pinpoint any weak points where an incident could occur. In case of spills, collection drains, pumps, and/or wet vacuums may be required to collect washdown waters to prevent discharge into waterways. Fire extinguishers should be placed in all storage areas.

PREVENTIVE MAINTENANCE

Preventive maintenance involves inspection and testing of facility equipment and systems to uncover conditions that could cause breakdowns or failures that may be prevented by adjustment, repair, or replacement of items. A good preventive maintenance program should include: (1) identification of equipment and systems that require preventive maintenance; (2) periodic testing and inspection; (3) appropriate adjustment, repair, or replacement of items; and (4) complete preventive maintenance records.

Equipment described in Section 4 should be checked to ensure that proper controls, alarms, and other precautionary features are in place. This includes detection equipment to pinpoint any leaks before they become massive spills or releases. Atmospheric monitoring equipment that could be utilized is identified in Table 5.

The life span of equipment (such as internal steam coils) should be estimated based on previous data and experience, and replacements should be made before the life-span limit. To reduce failure from corrosive action, prolong life, and reduce replacement costs, the temperature and environment should be considered in material selection. Recordkeeping is of extreme importance in keeping track of the age and life span of components so that adjustment, repair or replacement can be done at the proper time.

INSPECTION SYSTEMS

The purpose of an inspection system is to detect actual or potential incidents using visual observation, non-destructive testing, acoustic monitoring and other suitable methods. Inspection procedures should be written and the frequency of inspection determined. Records identifying the completion date and results of each inspection should be maintained for at least three years and signed by the appropriate supervisor. Establishment of the inspection recording system should ensure that adequate response and corrective action has been taken. This recordkeeping system can be incorporated into the facility spill reporting system.

Three major types of inspection should be performed: pre-service, in-service (periodic), and equipment termination. Pre-service inspection requires that new equipment received by the facility undergo inspection to ensure that the equipment is made according to design and construction specifications and is not faulty. In-service or periodic inspections are conducted during the operational phase of the equipment. This type of inspection throughout the expected life of the equipment will indicate possible malfunctions. The termination inspection occurs before equipment is taken out of service.

SECURITY

A security system is necessary to prevent accidental or intentional entry to a facility that may result in vandalism, theft, sabotage, or other improper or illegal use of plant facilities that could possibly cause a release. Elements of security to be considered for the SPP plan include:

- Routine patrols of plant security personnel
- Fencing
- Good lighting

TABLE 5. ATMOSPHERIC HAZARD GUIDELINES

Monitoring Equipment	Hazard	Ambient Level	Action
Combustible gas indicator	Explosive atmosphere	<10% LEL*	Continue investigation.
		10%-20% LEL	Continue on-site monitoring with extreme caution as higher levels are encountered.
		>20% LEL	Explosion hazard; withdraw from area immediately.
Oxygen concentration meter	Oxygen	<19.5%	Monitor wearing SCBA.*** NOTE: Combustible gas readings are not valid in atmospheres with <19.5% oxygen.
		19.5%-22%	Continue investigation with caution SCBA not needed, based on oxygen content alone.
		>22.0%	Discontinue inspection; fire hazard potential. Consult specialist.
Radiation survey	Radiation	<3 mR/hr**	Continue investigation. If radiation is detected above background levels, this signifies the presence of possible radiation sources; at this level, more thorough monitoring is advisable. Consult with a health physicist.
		>10 mR/hr	Potential radiation hazard; evacuate site. Continue monitoring only upon the advice of a health physicist.
Colorimetric tubes	Organic and inorganic vapors/gases	Depends on species	Consult standard reference manuals for air concentrations/toxicity data.

* LEL = Lower Explosive Limit
** mR/hr = milli Roentgen/hr
*** SCBA = Self Contained Breathing Apparatus

TABLE 5. ATMOSPHERIC HAZARD GUIDELINES (continued)

Monitoring Equipment	Hazard	Ambient Level	Action
HNU photoionizer	Organic vapors/gases	1) Depends on species	Consult standard reference manuals for air concentrations/toxicity data.
		2) Total response mode	Consult EPA Standard Operating Procedures. Cross readings above 500 ppm require maximum level of protection.
Organic vapor analyzer	Organic vapors/gases	1) Depends on species	Consult standard reference manuals for air concentrations/toxicity data.
		2) Total response mode	Consult EPA Standard Operating Procedures.

- o Traffic control
- o Controlled access at main gate entrance
- o Visitor passes
- o Locked/guarded entrances (on and off hours)
- o Locks on certain drain valves and pump starters
- o Television monitoring (not necessary for small facilities)

The existing security system should be evaluated to ensure that those areas or equipment identified as having the potential for incidents are adequately monitored. Based on this evaluation, recommendations for changes should be directed to facility management. Security measures include the use of security personnel for leak surveillance.

EMPLOYEE TRAINING

Employee awareness is the key to an effective spill prevention and control program. Spill prevention training can be carried out as part of a general training program or it can be done by itself. New employees always should be briefed on this important subject. Employee training meetings should be conducted at least annually to ensure adequate understanding of the objectives of a spill prevention program. The meetings should highlight previous spill events or failures, malfunctioning equipment components, and recently developed precautionary measures.

Adequate training in a particular job and process operation is essential for understanding potential release problems. Training sessions also should include notification procedures, spill prevention and control procedures, and emergency response actions.

Environmental incident or spill drills serve to improve employee reactions to releases. Well-planned drills should be conducted at least semi-annually to pinpoint important potential spill response problems in the facility.

Area procedures and training protocol should be reviewed annually by the SPP Committee and updated and/or improved. Area spill prevention reviews should be made periodically at frequencies suited to the size, complexity and nature of the area. The review should include:

- o process development and operations
- o plant projects and modifications
- o startup and shutdown preplanning
- o activities during turnarounds
- o spill amelioration

o communications

o spill reporting

o field inspection of potential spill areas

6. Preventive Engineering Practices

Preventive engineering practices (PEPs) may be thought of as SPPs that are oriented toward equipment rather than procedures. They are specific to groups of toxic and hazardous substances and to the potential sources of spill, that is: storage areas; loading/unloading areas; inplant transfer, process and materials-handling areas; drainage from plant site and secondary containment structures; and waste storage, treatment and disposal facilities.

PEPs are divided into pre-release and post-release groups, both designed to confine release within the facility boundaries. The difference between the two is that pre-release PEPs are of a general precautionary nature, whereas post-release PEPs are activated as spill control devices by a release. Typical pre-release PEPs include monitoring and alarm systems, non-destructive testing, labeling all storage, process and flow equipment, and proper storage procedures. Typical post-release PEPs include secondary containment of liquids and solids by dikes and berms, flow diversion, vapor control, and dust control.

The following subsections are grouped by equipment categories, emphasizing components found in chemical manufacturing facilities.

BULK STORAGE

Bulk storage consists of aboveground vertical cylindrical tanks with various roof configurations, aboveground horizontal tanks, and underground storage tanks. Many of these storage tanks are pressure vessels. A pressure vessel is usually considered to be a closed container of a fluid under pressure used to perform some process function. Storage tanks, heat exchangers, evaporators and reactors are examples of pressure vessels. Vessel design requires a thorough knowledge of the code requirements, the properties of the fluid being contained, the properties of suitable materials of construction, and the methods for fabrication using these materials. Uncontained bulk storage under cover is also used in chemical manufacturing facilities, however, open stockpiling should be avoided.

Tank Construction

Bulk storage tanks can be classified by several different techniques, such as, the product contained, the method of construction, the general purpose, and the general configuration. In terms of configuration, storage tanks can be found in a number of different sizes and shapes. The basic aboveground, vertical storage tank is a cylinder made from welded steel plate. These tanks can generally withstand pressures from 0 to 0.5 psig. The most common roof for these tanks is the fixed cone roof. The roof is supported by a center post resting on the tank bottom and a system of steel rafters extending from the center post to the tank shell. In this typical storage tank configuration there is an ullage space above the liquid that contains a vapor/air mixture. Venting devices are provided to prevent sizable pressure changes during tank filling or unloading operations.

Since control of the vapor/air ullage space in the tank is of critical economic and environmental importance, the floating roof tank and others have been developed. The roof of the floating roof tank floats on top of the product, riding up and down inside the tank shell as the liquid level changes. Special reinforcement must be installed around the circumference of the tank near the top to provide support for the side wall. Flexible "shoes" are used to close the space between the roof and sidewall to prevent escape of the product. "Pontoon" seals have also been developed to add stability to the sealing capability. In addition, the floating roof can be double-decked to prevent product exposure to roof skin temperatures. The floating roof essentially eliminates vapor discharge when adding a product, and air uptake when removing a product. As a result, air-induced corrosion is greatly reduced. A disadvantage is that heavy rainfall or snow accumulation can force the roof to sink into the product, causing a release.

Another tank configuration is the outside breather roof storage tank. This tank utilizes a modification of the basic cone roof designed so that when the roof is resting on its supports, it is in the form of an inverted cone. When a liquid product level is low, the tank roof drops and rests on the supports. As a product is added or as the temperature inside increases, the roof rises above the supports. The advantage of this system is the elimination of vapor loss to the environment, as long as the tank is kept nearly full and the vapor created does not exceed the volume provided by the roof displacement. If excess pressure does occur, it is relieved via a control valve. A disadvantage is that the roof metal actually bends as pressure changes occur, potentially leading to metal fatigue and structural failure.

Another example of tank configuration is the outside lifter roof storage tank. In this case, the roof is an entirely separate member from the shell, and it moves within guides a vertical distance of several feet. This permits a larger vapor space than the previous designs, yet does not subject the metal structure to bending. Vapor release is prevented through either a liquid seal by having a metal skirt extend downward from the roof into a liquid filled trough built on the upper part of the tank shell, or a flexible curtain of treated fabric, impermeable to product vapors, that is attached to both the removable roof and the stationary tank shell. Weather protection is necessary for both types of seals.

Other tank arrangements include tanks for outside horizontal storage and for underground storage. Tanks of the former type are utilized for storing both liquids and gases in both pressurized and nonpressurized modes. Pressurized tanks are identified by the elliptical or hemispherical heads which provide greater structural integrity. Relief valves are similar to those provided on the vertical tanks. Underground storage tanks include remote shutoff switches for shutting down the pump should the excess flow valve not close. Tank vents prevent this buildup of pressure when filling the tanks and will overflow the product should the tank be filled too much. A significant problem with underground storage is that of corrosion. This has been overcome somewhat by the use of fiberglass tanks which resist leakage to a great extent. Small storage vessels can include gas cylinders for welding or heating, 55 gallon drums and other pressurized and nonpressurized vessels, the latter not having any relief valves and/or fusible plugs to prevent pressure buildup.

The minimum construction requirements for design, fabrication, inspection, and certification of unfired pressure vessels are detailed in Section III of the American Society of Mechanical Engineers (ASME) Boiler and Pressure Vessel Code. The design pressure is customarily set at 15 psi or 10%, whichever is greater, over the operating pressure for vessels without cyclic swings, while for vessels with cyclic pressure variations the design pressure is usually 5 to 10% greater than the highest anticipated pressure. The design temperature is usually 50°F above the maximum temperature that will not result in a decrease of the code - allowable stress that would cause an increase in thickness. Some materials of construction also have specific limitations on temperature.

Pressure vessels may be fabricated by drafting, riveting, forging, casting, brazing, welding, or a combination of these. With the exception of a few special cases, practically all industrial pressure vessels are of welded construction. Section IX (Welding and Brazing Qualifications) of the ASME Boiler and Pressure Vessel Code details the requirements for qualification of procedure and operators.

Tank Support

An important factor related to tank integrity is proper tank support. The design of supports may be quite complex due to secondary stresses, moments, and shears caused by support attachments. Tank supports must be built on a sound, well-compacted base. A shift in the support or underlying ground can create stress in the metal that may lead to leaks or structural failure. Vertical vessels may be supported by legs, lugs, or brackets, either with or without a ring girder, or a skirt may be used. Large horizontal tanks are preferably supported by not more than two saddle supports. Also, the structural supports for horizontal tanks can be weakened during a spill fire, causing the tank to drop and rupture. Steel supports for any flammable liquid tank should be protected by fire resistive material with a two-hour rating. Saddles made of concrete maintain their integrity for a longer time than unprotected steel. While the danger of collapse is eliminated when storage tanks are placed directly on the ground, corrosion of the metal in contact with moist ground is a problem. Also, the vessel and its supports must be designed to accommodate loadings resulting from the weight of the vessel and its contents, externals (e.g., ladders, platforms and piping), insulation and fireproofing, and wind and earthquake.

Tank Placement

Plant geography should be studied in planning the location of storage vessels. Ground elevations will affect drainage patterns. Seepage into the water table must be prevented by locating storage tanks where water-table depth is substantial and soil porosity is limited. Close proximity to the storage vessel of storm drains and watercourses, i.e., rivers, creeks or ditches that may contain water during thaw or rainfall conditions, is to be avoided. Provision should be made in an emergency to shut the drains or to bypass the flow through a retention basin that might hold the spill. Tankage should also be located in a downwind direction away from traffic corridors and residential areas in a secluded section of the plant to enhance its identification as a hazardous area and to minimize extraneous traffic through the area. Accessibility into the storage area and approach from all sides must be provided for emergency operations. Exposure of other tanks, buildings or equipment to a storage vessel release should be limited. Adequate water supply, such as hydrants and static water sources for firefighting purposes, should also be available.

Tank Monitoring

Non-destructive examination of boiler and pressure vessels is regulated by Section V of the ASME Boiler and Pressure Vessel Code. Political subdivisions are free to adopt this code which is designed to promote greater safety to life and property through uniform inspection practices and consistency in the construction, installation, and inspection of safe boilers and pressure vessels.

Non-destructive, tank integrity testing programs can be done either by plant personnel directly or through outside contracting arrangements. Tank shells should be tested on at least an annual basis using ultrasonic devices to accurately measure the wall thickness. Five year testing programs for lower heads or tank bottoms are also employed in conjuction with tank cleaning operations that permit entry into the tank.

Rubber-lined tanks designed for special material containment are often spark tested to verify liner integrity. This special purpose testing is generally provided by contracted personnel on a two year interval basis.

Epoxy-resin coatings of the internal surfaces of bulk storage containers is coming more into use. A typical approach is to sand-blast the internal surfaces down to bright metal, thereafter applying two coats of epoxy to the internal metal surface. The epoxy fills the pits in the metal, thus retarding corrosion that might lead to pinhole leakage. New tankage is now engineered to carefully match probable business and product storage needs.

Material Storage

The open stockpiling of ores, chemicals, and minerals should be discontinued. Piles of bulk material should be covered to prevent leaching and runoff. If open shed-type structures are used for bulk storage, perimeter curbing should be provided or perimeter drainage trenches should direct runoff into a suitable treatment facility. Bulk storage should be located on pads of concrete or other relatively impervious material to prevent ground water leaching and percolation into the earth.

Metal and fiber containers should be stored in a covered area, off the ground. The area should be provided with drainage to a treatment facility similar to that provided for diked storage tanks. Containers should be segregated so that leakage or releases from them will not spread through the operating area and pose an incompatibility hazard.

Space should be provided for the temporary storage of raw materials at process points and for finished products in the manufacturing area. This will permit continued operation during abnormal situations in continuous operations and alleviate safety and operational problems in batch operations.

Alarms

Some alarms are designed for operating convenience, others warn of approaching catastrophe. High and low liquid level alarms are quite commonplace. Recent technological advances have produced reliable alarms, many of which are connected directly to pump shut-off controls for protection against overfill. Where sequence controllers are used, there should be an automatic check, together with alarms, at key steps after the controller has called for a change. There should also be a check together with alarms at key steps before the next sequence changes.

Pilot lights are similar to alarms in function. There are those whose failures to function properly may cause injury to people or major loss of money, and those that are simply for operating convenience. It is important to be sure that the listed condition for indication is the critical variable rather than a dependent variable. Warning lights are also provided as a fail-safe precaution in loading/unloading operations (Section 6).

Fire Protection Systems

Automatic sprinkler or automatic water spray (deluge) protection should be provided wherever there are combustible materials stored. Suitable drainage for spilled flammable liquids and fire protection water from sprinklers and hose streams should also be intalled. Sprinkler waterflow alarms should be arranged so that any flow of water through the sprinkler piping will give an audible local alarm and automatically transmit waterflow signals to a central supervised location.

Sufficient hydrants having two outlets should be available to provide two - 2.5 inch hose streams at any point in the property where fire may occur regardless of wind direction. This will require spacing hydrants 225-250 feet apart at plants with ordinary hazards. At plants having highly combustible occupancies, the spacing may need to be reduced to 100-150 ft. For average conditions hydrants should be located 50 ft from the building or equipment protected.

A loop system with sectional valves is recommended for extension of underground fire mains because it assures greater reliability by providing water flow from two or more directions. The mains should be of ample size to carry the maximum fire water demand for sprinklers, hose streams, and monitors. Sectional control valves in the underground fire mains should divide the grid system into sections and thus limit the area subject to a single impairment. With any one section shut off, at least one of the water supplies should be available on the remainder of the system.

Fire protection monitors are not a substitute for automatic sprinkler protection or hose streams. They should be used to protect equipment containing flammable liquids, not housed in a building or structure protected by automatic sprinklers. Their principal advantage is to provide a quick stream of water while hose lines are being laid, particularly where work force is a problem.

Fireproofing of all main load-bearing structural members that support either process piping or equipment within hazardous areas should be done. Firewalls, partitions, or barricades should be used to separate: (1) important property values, (2) units important to continuity of production, and (3) high hazard operations. Also fire protection for indoor and outdoor storage of flammable liquids in drums is recommended.

Special hazard protection including fixed carbon dioxide, foam, dry chemical, and explosion suppression systems should be employed where needed. Fire detectors should be provided where automatic sprinklers cannot be justified due to the limited quantity of combustibles or the water reactive nature of the stored materials. Also, fire extinguishing agents must be compatible with process materials. Flame arresters can be installed in the vapor space of storage tanks to lower the temperature of the vapor below its auto-ignition point. Flame arresters are usually metal tubes, but metal plates and screens are also used.

Bonding and grounding should be provided to protect against static electricity buildup. Tank grounding should be coordinated with cathodic protection, and all structural steel should be grounded.

Secondary Containment

Spillage at a fixed facility can occur from any one or more of several causes, such as: overfilling storage tanks, leaks from structural stress, leaks from valves or piping, or tank failure from a natural catastrophe. Because tank farms are usually located in industrialized areas where adequate room is not available for impoundment basins, the solution is to build a dike around individual tanks or groups of tanks. This wall would contain the spill and keep it from spreading. Dikes are now engineered structures rather than bulldozed earthen structures. The materials used for the dikes can be earth, steel, or concrete. The bases are engineered to provide full "water-tight" enclosures generally using impermeable membranes in a masonry foundation. Earthen dikes are designed to a height of 6 ft with a 1:2 slope and are gunnite coated over reinforcing wire. The dikes must be strong enough to withstand the head pressure created by the released liquid. The size of the dikes is dependent upon the type of product contained in the tanks. Products that are not subject to boilover should have sufficient diked area to contain the volume of the largest tank. Dikes for products subject to boilover are usually designed to contain the volume of all the tanks enclosed. Deflectors are added to the top of the dike to prevent the liquid wave from washing over the top in case of a sudden release. Furthermore, large dikes can be used to enclose several tanks while the enclosed area can be subdivided by low walls into smaller units so that minor spills will not needlessly expose other tanks.

Dikes must be constructed to contain not only a released product but also rainwater and water that may accumulate from firefighting operations. To counteract water buildup, dikes should be equipped with control-valved drains (Section 6). Also, if the dike wall is too high, access for fire fighters or egress for plant workers may be difficult or impossible to achieve.

LOADING AND UNLOADING AREAS

Areas included in this category include rail switch tracks and rail tank car loading racks; tank truck loading racks, both overhead and bottom loading; building loading platforms or docks; marine docks; pipeline connections; valve manifolds; and any other areas where a product is loaded or unloaded in bulk quantities or drum containers. Loading/unloading facilities should be located downwind and in a peripheral location from the main plant area, and traffic to these facilities should not pass through the plant. Secondary containment must be provided so that a spill will not leave the property, even in adverse weather. (Section 6).

Tank Truck Loading/Unloading

A system of containment curbs should be used for tank truck loading/unloading areas. Ramps should be provided to allow for truck access into confined areas. The curb enclosure should be designed to hold at least the maximum capacity of any single tank truck to be loaded or unloaded in the plant. Loading rack areas are now most often concrete-surfaced with drainage directed to one or more holding tanks equipped with pumps to recover separable materials. Also, tight vapor emission control has dictated the design of new loading equipment to eliminate or drastically reduce the emission of vapor during loading operations. Two types of containment systems are typically employed: a quick drainage system or diversionary structure which diverts flow to a sump/recycling network or a dike, catchment basin, containment area. A quick drainage system is frequently employed on new installations where it can be built in at the outset. The drains are typically connected to an underground tank or sump of a volume sufficient to contain the contents of the largest truck using the plant. The advantages of such a system are: cleanliness, easy maintenance, and minimum use of aboveground space. A roof can also be constructed over the curbed enclosure to keep storm runoff out of the collection system. A diversionary structure system can be constructed to divert flow to another location where the spilled substance can be collected, contained, and removed prior to a release from the facility. The rack loading site can be graded and curbed to divert the runoff into a diked area several feet lower in elevation. The diked area can also be used to contain the storage tanks. Drain valves should be kept locked in the closed position.

As a fail-safe precaution, an interlocked warning light or signs and a physical barrier system should be provided in loading/unloading areas to prevent vehicular departure before complete disconnect of the transfer lines. Instructions should also specify that loading and unloading will be accomplished only while an operator is in attendance at the loading rack. Prior to filling and departure of any tank truck, the drains and outlets of the vehicle should be closely examined for leakage, and if necessary, tightened and adjusted to prevent leakage while in transit. Electric controls for pumps should be locked individually or located within a locked cabinet to prevent the pumps from being activated by unauthorized persons.

Railroad Tank Car Loading/Unloading

A trenching system should encompass each railroad tank car unloading area. The trench should be designed to direct any spill to a catchment basin or holding pond, at least large enough to contain the largest tank car loaded or unloaded in the plant. Modern improvements call for installation of drain pans beneath rail loading/unloading facilities, plus the use of quick shutoff fittings on hoses and piping.

Static grounding should be provided on flammable liquid tank cars (and tank truck) loading and unloading docks. Also, approved loading platforms, including handrails, should be provided to work areas of tank cars, hopper cars, and trucks.

Marine Loading/Unloading

Marine loading docks have been improved over the years to the point that deck surfaces are now constructed of concrete with spill containment curbs and engineered drainage directed back to a "slop tank". Drainage valves are employed but they are subject to supervised opening for release of uncontaminated rainwater. All ships and barges loading or unloading hazardous "floaters" should be effectively boomed in the area of the transfer operations. Transfer pumps and flange connections should be positioned on shore and confined within a suitable containment curb. Catch trays should be positioned under potential leak areas. Pressure-drop alarm and shut-off systems should be provided on the transfer lines to minimize losses should a break in a transfer line occur. All pump controls on marine docks should be secured in the closed position or electrically isolated when not in service. For the prevention of spills during flexible hoseline connection and disconnection, positive closure butterfly valves should be installed immediately adjacent to the terminal flange connection. These valves permit manual opening and closing of the hoseline to retain a product that cannot be drained from the line after product transfer. Additionally, to prevent deck level spills, a drainline between the check valve and the vessel's manifold valve can be provided to drain the contents of the line back into the vessel's cargo compartment prior to line disconnect.

IN-PLANT PROCESS AND TRANSFER

Based on experience, an acceptable plot plan for process units may consist of a long, straight, "in-line" arrangement of the towers, drums, exchangers, pumps, and main pipe alley. Traversing the in-line arrangement, one might consider a roadway; a gantry way with cooling water headers underneath; a line of fractionating towers, heat exchangers, accumulators, reflex drums, etc. with the grade below this equipment depressed about 8 inches; the main pipe alley; a row of pumps; and possibly another roadway.

A minimum of two means of access for the site should be provided to facilitate entrance by emergency vehicles and not more than one of these should be subject to uncontrolled blockage, i.e., railroads, highways. Also, proper banking, barricades, and warning signs should be in place where sharp curves are a necessity in access roads. (In areas where fog conditions may obscure visibility, provide rumble strips to warn of an approaching curve.)

Also, boiler plants, maintenance shops, and warehouses (with railroad spurs) are potential ignition sources and require an upwind location and separation by reasonable distance from areas of leak hazard.

Process Materials/Equipment

Process materials that are unstable, pyrophoric, or spontaneously ignitable must be controlled. Available data on the amount and rate of heat evolution during decomposition of all materials in the process should be utilized. Storage and piping systems for flammable materials must receive special attention. Proper safeguards for flammable dust hazards must be provided. Materials of construction must be compatible with the chemical process materials that are involved. A reliable inert gas supply for purging, blanketing, or inerting, including a true sweep-through purge for startup and shutdown, must be provided. (Reference: National Fire Protection Association (NFPA) Standard No. 69 for Explosion Prevention Systems.) Provision should be made for rapid disposal of reactants if required under emergency conditions. Gradual or sudden blockage in equipment, including lines, should be prevented. Protection should also be provided against extreme weather conditions that may adversely affect raw materials or process materials or conditions.

In any high hazard operation, explosion suppression equipment may be needed to stop an explosion once started. Process equipment made of glass or other fragile material should not be used. If a more durable material cannot be used, the fragile material must be adequately protected to minimize breakage. Where sight glasses are necessary on process equipment that is subject to hazardous conditions (flammables, toxic, high pressure, temperature extremes), pressure equipment should be used. Pertinent equipment, especially process vessels, should be checked for pressure capability. Requirements should be reviewed for concrete bulkheads or barricades to isolate highly sensitive equipment and protect adjacent areas from disruption of operations. Manually operated valves, switches, and other controls should be readily accessible to the operator from a safe location. Also, free-swinging material hoists should not be installed.

Interlocks are often used with process equipment. An interlock is an arrangement of equipment that ensures the existence of certain specific conditions before an action can be taken or for an action to continue. Some interlocks are designed for safety purposes to prevent major accidents.

Instrumentation

Suitable process control and safety instruments should be utilized based on functionality. In the event that all process and safety instruments should fail simultaneously, the collective operation should be fail-safe. Instruments directly or indirectly significant to process control should exhibit minimal response time lag. Highly exothermic reactions should be protected by dual, independent instrumentation, including alarms and interlocks.

Gauges, meters, and recorders should be designed and installed in such a manner that they can be read easily. Instruments which cannot be read easily should be replaced. Extremes of atmospheric humidity and temperature should not affect instrumentation. Instrument packages should be properly installed, grounded, and designed for the environment. Procedures must be established or in effect to test and prove instrument functions, including interlocks and safe action in the event of power or instrument air failure. Instrument performance and potential malfunction should be assured by periodic testing.

It may be necessary to ensure that instrument power be available on loss of plant power. Also, quality of power, i.e., voltage or frequency variations not greater or less than 5%, may be important for some instrumented processes. The electrical system should be isolated physically from process systems as much as possible to minimize exposure to fire, corrosion, or mechanical damage. Also, emergency lighting should be provided for escape or emergency shutdown operation in the event of power failure.

Piping

Piping systems should be properly supported and guided. Stresses and movement due to thermal expansion should be minimized. In this regard, piping systems, in particular, instrument connections and lines in dead-end service, should be provided with antifreezing protection.

Buried pipelines should be avoided since corrosion can produce leakage that can remain undetected for long periods of time. Lines that pass under roadways or other subsurface locations should be either trenched or installed in conduit to facilitate inspection and replacement, if required.

However, the number of road crossings should be minimized. Trenches for pipe alleys also should be avoided since trenches may transport leaking flammables or toxics from one area to another. Where pipe alleys cross above roadways, overhead clearance must allow for easy passage of heavy equipment such as cranes. Also, pipe alleys should not pass through diked areas. A fire in a diked area can rupture pipes. External coating and protective wrapping of piping, especially buried installations that cannot be avoided, has become commonplace.

Pipelines into a facility and pipe alleys within are low-potential sources of spills. However, they should be protected if possible by diking when above ground. The diking can serve a dual role as (1) containment for a spill if it occurs, and (2) protection against damage from contact. The piping should be protected cathodically against electrolytic corrosion and should be pressure tested periodically to ensure integrity. When piping is no longer in use, it should be emptied of product, blown with an inert gas and sealed shut with a blind or blank flange. Openings for pipes through walls or floors, as well as holes caused by removal of old equipment and piping, must be properly caulked to prevent leaks if they present potential exit points of a spill. Piping supports should be adequate and flexibility should be incorporated to keep forces on machinery (due to thermal expansion of piping) within acceptable limits. Wood pipeline supports should be avoided since wood can retain moisture and cause pipeline corrosion which, when coupled with the pulsating action of the line, could cause line failure with resulting leakage. Supports should be designed to allow for the pulsating movement of the line with only a limited point of surface contact.

Valving

Each product fill-line which enters a tank below the liquid level should have a check valve located close to the bulk storage tank. The check valve confines the product to the tank in the event of valve or pipeline failure. It also permits overhaul of the main shut-off valve and prevents shock loading of the pipeline and valves from a "slug" of the tank contents caused by backflow into an empty fill line. Check valves should be positive and fast-acting to prevent reverse flow and reverse rotation of pumps, compressors, and drivers.

Small leaks from fittings, valve packing glands, bearings, and seals should be accumulated through a floor drain into a sump or by the use of drip pans, sorbents, or other means. The sump should have a capacity sufficient to handle a surge of product or water. Pumps used to transfer sump contents should not produce emulsions or dispersions.

Cast iron valves should not be installed in piping subjected to strains or shock service. Nonrising stem valves should be avoided wherever possible. Also, the use of double block and bleed valves on emergency interconnections to avoid cross-contamination should be encouraged.

Venting

Manifold venting systems should be avoided. Venting system design should include provisions for removal, inspection and replacement of relief valves and rupture disks. Flame arresters should not normally be installed on relief valve discharges, or rupture disks on pressurized vessels. Sizing information on emergency relief devices, such as, breather vents, relief valves, rupture disks, flame arresters, and liquid seals, can be obtained from S&PP Briefs 42 and 100. Where rupture disks are used to prevent explosion damage, they must be sized relative to vessel capacity and design. Where rupture disks have delivery lines to or from the disks, adequate line size relative to desired relieving dynamics must be assured.

Equipment vents should terminate outside buildings and structures and located to prevent injury to personnel or damage to property in the event of discharge. Discharge piping should be supported independently of relief valves; be as short as possible, and have minimum changes in direction. Piping also should be supported properly to prevent bending or whipping upon relief discharge. Drain connections or weep holes in discharge piping of relief valves and relief valves should be provided where condensate, rain, or snow could collect. Weep holes should not be permitted to discharge flammable gases or liquids in such a way as to expose tank shells to fire in the event of ignition during operation of relief valves. Low-inertia weather covers at the end of the discharge line and liquid disengagement, where necessary, should be provided. Relief valves should be provided on the discharge side of positive displacement pumps, between the positive displacement compressor and block valves, between back-pressure turbine exhaust flanges and block valves, and on any equipment where liquid can be trapped and later warmed.

Where rupture disks are in series with relief valves, valve corrosion or toxic material leakage can be prevented by installing a rupture disk next to the vessel and monitoring the section of pipe between the disk and relief valve with a pressure gauge and pressure bleedoff line. Valves should not be placed in lines to the tee, pressure gauge, or bleedoff. Flow checks are required in bleedoff lines. Provisions should be made for maintaining piping to relief valves and vacuum breakers at a proper temperature to prevent any accumulation of solids from interfering with the action of the safety device. Assurance should be made that the service temperature at which the rupture disk bursting pressure is stated by the supplier is actually the rupture disk operating temperature.

Color Coding/Labeling

All piping should be color coded. Each product pipeline should be clearly marked by lettering, color banding, or complete color coding to indicate the product within. Dedicated piping can show the name of the product on the line. Some plants install fabric covers of a standard color over flange joints that, in the event of product leakage, change color.

DRAINAGE CONTROL

Surface drainage within the confines of a plant should be engineered so that spills of hazardous substances can be effectively contained. Two separate drainage systems, one handling surface runoff rainwater, the other draining production water or water that could possibly contain pollutants, are recommended. The trenches should be concrete (gunnite) lined with, in the latter system, specially designed sumps to retain floatable materials for recovery purposes. Rainfall should be properly directed away from high-potential spill areas. Drainage around storage tanks should be graded so that a spill will normally flow to a holding pond capable of containing the contents of one or more tanks.

Facility Diking

Facility perimeter containment has also been employed whereby the boundary land area is raised to develop a "saucer-shaped" enclosure to contain accidental releases onto plant property. The final discharge of all in-plant drainage should be equipped with a diversion system that could, in the event of an uncontrolled release, be returned to the plant for treatment via a closed-loop system.

Road Drainage

In-plant roads should be paved and guttered. If a spill on the roadway could reach a watercourse via storm drains, provision must be made to shut the drains in an emergency, or to bypass the flow to a retention basin that would hold the spill. Proper drainage should also be provided in the case of leakage in any in-plant drum storage area, such as via a concrete pad drained through a containment barrier, or along the path by which drums are transported from filling to storage to shipping.

Plant Drainage

Only essential plant floor drains should remain open for use. Unnecessary drains should be closed. Areas where floor drains are deemed essential should be subdivided by the use of curbing to segregate water drains from possible contact with a released product.

Drainage Valving

Water drawdown valves used to drain accumulations of condensate water and other contaminants from the lower portions of a storage tank should not drain directly into diked enclosures but should be piped to the plant wastewater treatment facility. Drainage from diked storage areas should be valved to prevent a product spill from entering the drainage discharge or, in the case of excessive product leakage, the in-plant waste treatment system. Valves used for drainage of diked areas should, as far as practical, be of manual, open-and-close design.

Secondary Containment

The condition of the retained rainwater should be determined before drainage of impounded waters into off-site watercourses or storm drains is allowed. All plant drainage systems should flow into retention ponds, lagoons, or catchment basins. Segregation of these retention ponds or catchment basins to prevent commingling of incompatible released substances is required. Such ponds are often lined with concrete, synthetic membranes or impermeable clay, and contain baffles or weirs designed to prevent accidental escape of a product in the event of massive product discharge. Test wells are often drilled to monitor groundwater conditions indicative of leaching from these ponds.

WASTE STORAGE, TREATMENT AND/OR DISPOSAL

Drainage from production areas should flow directly into the plant's waste treatment ponds or, if acceptable to the municipality, into the nearest government-operated waste treatment facility. Recently some facilities have established, where feasible, waste treatment or "shot" ponds to enable emergency holding for at least one and preferably two days of plant processing waters. Retention allows for subsequent treatment and the natural gravity separation aides spill recovery, all of which is part of contaminated water spill control.

The handling of a product collected in the maintenance of building and machinery must be properly controlled to prevent product discharge. Cleanup water containing toxic chemicals cannot be dumped down the most convenient plant flow drain. Specific disposal areas should be provided for cleanup solutions, solvents and the like. Also, returned drums should be inspected and stored on a drained pad, and roofed to minimize rainwater accumulation within a product containment dike.

It is assumed that, for specific substances requiring neutralization or countermeasures of some type if spilled, a plant will have proper chemicals to carry out such counteraction. Each generator, transporter, or storer should ascertain approved chemical treatment agents, and an adequate supply of such agents should be stored at the plant for the treatment of chemical spills. These chemical agents cover neutralizers, gelling agents, firefighting foams, dispersants, and other materials. Whenever acids are maintained in bulk storage, an adequate supply of neutralizing material, such as caustic or lime, should be available on the plant property. The neutralizing agent should be stored as closely as practical to the acid storage area and in sufficient quantity to neutralize the contents of the largest capacity of acid contained in any one storage tank. Where drainage waters are chemically treated in more than one treatment unit, natural hydraulic flow should be used. If pump transfer is needed, two "lift" pumps should be provided, and at least one of the pumps should be permanently installed.

A supply of personal protective safety equipment, such as rubberized cover-alls, rubber boots, safety goggles, self-contained breathing apparatus, gas masks, and rubber gloves should be maintained for immediate use in convenient spill control locations and should be compatible with potential applications. This supply should be checked periodically to see that the equipment is in good order and useable.

When non-miscible chemical floaters are handled and stored within the plant, an adequate length of flotation spill containment boom or an adequate amount of sorbent material should be available with suitable means to position the boom strategically or sorbent satisfactorily, and thus confine the spilled material. Sorbents are useful tools for the control of releases of hazardous liquids. Sorbents are classed as natural organic, natural inorganic, and synthetic. Most commercial sorbents are readily available, fairly light in weight, easy to handle, and require only moderately careful storage. Generally, for disposal, saturated sorbent is put into drums, plastic bags, or occasionally, open trucks for disposal either by incineration or by burial in a permitted landfill.

7. Bibliography

1. Cleary, J. G., O. D. Ivins, G. J. Kehrberger, C. P. Ryan and C. W. Stuewe, 1979. NPDES Best Management Practices Guidance Document. Hydroscience, Inc., EPA Contract 68-03-2568, Report EPA-600/9-79-045.

2. E.I. DuPont de Nemours & Co., Inc., 1973. Spill Prevention Control Program, Beaumont Works Environmental Control Department, TX.

3. EPA Office of Water Enforcement, Permits Division, 1980. NPDES Best Management Practices Workshop, Chantilly, VA.

4. Fawcett, H. H. and W. S. Wood, Editors, 1982. Safety and Accident Prevention in Chemical Operations. Second Edition, John Wiley and Sons, New York, NY.

5. Midwest Research Institute, 1980. Cost Analysis for Proposed Hazardous Substance Pollution Prevention Regulation, Final Report. EPA Contract 68-01-3861, Kansas City, MO.

6. Midwest Research Institute, 1982. Cost Analysis for Hazardous Substance Pollution Prevention Regulation, Draft Final Report. EPA Contract 68-03-3014, Kansas City, MO.

7. Pace Co. Consultants and Engineers, 1975. Oil Spill Prevention Control and Countermeasure Plan Review. EPA Training Grant T-900-175-02-2 to Rice University, Houston, TX.

8. Perry, R. H. and C. H. Chilton, Editors, 1973. Chemical Engineers' Handbook, Fifth Edition. McGraw Hill Book Co., New York, NY.

9. State of California, 1974. Oil Spill Contingency Plan. Secretary of State, Sacramento, CA.

10. Trentacoste, N. P., G. R. Bierman, and J. Cunningham, 1980. Spill Prevention Control and Countermeasure Practices at Small Petroleum Facilities. Science Applications, Inc., EPA Contract 68-03-2032. Report EPA-600/7-80-004.

11. Unterberg, W. and R. M. Moorehead, 1980. Guide for Spill Prevention Control and Countermeasure Inspectors. Rockwell International, EPA Contract 68-03-2648, Newbury Park, CA.

Appendix: Fixed Facility Chemical Process Equipment Components

TRANSPORT AND STORAGE OF FLUIDS

Pumping of Liquids and Gases

Fluids can be made to move through a conduit or channel: (1) by action of centrifugal force, (2) by volumetric displacement, (3) by mechanical impulse, (4) by transfer of momentum, (5) by electromagnetic force, or (6) by gravity.

Centrifugal Pumps --
The centrifugal pump is the type most widely used in the chemical industry for transferring liquids of all types. They are available through a vast range of sizes; in capacities from 2 or 3 gal/min to 100,000 gal/min; and for discharge heads from a few feet up to several thousand feet. A centrifugal pump, in its simplest form consists of a impeller rotating within a casing. The impeller consists of a number of blades, either open or shrouded, mounted on a shaft that projects outside the casing. The casing consists of a chamber in which the impeller rotates, provided with inlet and outlet for the liquid being pumped. Casings are of three general types: circular, volute, and diffuser or turbine.

Process pumps are available in a variety of designs for particular services. The term is frequently applied to a single-stage, pedestal-mounted unit, usually with simple-suction impellers, designed for ease in dismantling, accessibility, and with packing built to handle corrosive or dirty liquids. Double-suction single-stage pumps are used for general water-supply and circulating service and for chemical service when handling liquids that are non-corrosive to iron or bronze.

Pumps with built-in electric motor or, sometimes, steam-turbine-driven are known as close-coupled pumps. Such units are extremely compact and are suitable for a variety of services. The canned-motor pump consists of a design in which the cavity housing the motor rotor and the pump casing are interconnected. As a result, the motor bearings run in the process liquid and all seals are eliminated. The in-line vertical pump uses a casing as an integral part of the piping system in similar fashion to a valve body. Other vertical pumps can be dry-pit mounted with the packing box located above the highest liquid level or wet-pit mounted with the column and casing actually immersed.

Sump pumps are small single-stage vertical pumps used to drain shallow pits or sumps. Multistage centrifugal pumps are generally used for services requiring higher heads than can be generated by single-stage pumps. Multistage pumps may be of the volute type or the diffuser type.

Propeller Pumps--
Axial flow (propeller) pumps are essentially very high capacity, low head units. Normally they are designed for flows in excess of 2000 gal/min against heads of 50 feet or less. They are used to great advantage in closed-loop circulation systems where the pump casing becomes an elbow in the line.

Turbine Pumps--
The term turbine pump is applied to units with mixed-flow (i.e., part axial and part centrifugal) impellers. Such units are available in capacities from 100 gal/min upward for heads up to about 100 feet per stage and are usually vertical. A common form of turbine pump has the pump element mounted at the bottom of a column that serves as a discharge pipe. Regenerative pumps employ a combination of mechanical impulse and centrifugal force to produce heads of several hundred feet at low volumes. These pumps are useful where it is required to handle low volumes of low viscosity, clean liquids at higher pressures than are normally associated with centrifugal pumps.

Positive-Displacement Pumps--
Positive-displacement pumps ideally produce whatever head is impressed upon them by the restrictions to flow on the discharge side. Overall efficiencies are higher than with centrifugal pumps because internal losses are minimized. One form of positive-displacement pump is the reciprocating pump. There are three classes of reciprocating pumps: piston pumps, plunge pumps, and diaphragm pumps. In general, the action of the liquid-transferring parts is the same, a cylindrical piston, plunger, or bucket, or round diaphragm being passed or flexed back and forth in a chamber. The device is equipped with valves for inlet and discharge of the liquid being pumped, and the operation of the valves is related to the motions of the piston. Piston pumps exist in two ordinary types, simplex double-acting and duplex double-acting. Simplex plunger pumps are used quite commonly as metering or proportioning pumps.

In rotary pumps, mechanical displacement of the liquid is produced by rotation of one or more members within a stationary housing. Rotary pumps are either interior bearing or exterior bearing. The former is lubricated by the liquid being pumped whereas the latter is oil-lubricated. Gear pumps use two or more impellers in a rotary-pump casing where the impellers in the form of toothed-gear wheels rotate with extremely small clearance between each other and the surface of the impeller and the casing. As the teeth of the impeller pass the suction opening, liquid is impounded between them and carried around the casing to the discharge opening. Screw pumps are a modification of the helical gear pump in which the liquid is fed by rotors to either the center or the ends, depending on the direction of rotation, and progresses axially in the cavities formed by the meshing threads or teeth. Screw pumps, because of multiple dams preventing slip, are well adapted for handling viscous liquids.

Fluid-displacement pumps employ displacement by a secondary fluid rather than mechanical action. One such device is the air lift, designed to raise liquid by means of compressed air. Another is the acid egg or blowcase consisting of an egg-shaped vessel that can be filled with a charge of liquid which is then forced out the discharge pipe by the admission of compressed air.

Jet Pumps--
Jet pumps are a class of liquid-handling devices that makes use of the momentum of one fluid to move another. Ejectors and injectors are the two types of jet pumps. The ejector, siphon, exhauster, or educator is designed for use where the head pumped against is low and less than the head of the

fluid used for pumping. The injector is a special type of jet pump, operated by steam, in which the fluid being pumped is discharged into a space under the same pressure as that of the steam used to operate the injector.

Electromagnetic Pumps--
All electromagnetic pumps utilize the motor principle--that of a conductor in a magnetic field and the directed force exerted on it when carrying a current. In these pumps, the fluid is the conductor and the force is exhibited as a pressure if the fluid is suitably contained. Both alternating current and direct current units are available. Multiple induction alternating current pumps exist in helical and linear configurations. Helical units are effective for relatively high heads and low flows, while linear induction pumps are best suited for large flows at moderate heads.

Fans--
Fans are used for low pressures (generally less than 0.5 lbs/sq in) and are usually classified as centrifugal or axial-flow type. Centrifugal fans are made in three general types, the straight-blade, or steel plate fan, the forward-curved blade fan, and the backward-curved blade fan. Straight-blade fans have rotors of comparatively large diameter with a few (5 to 12) radial blades, resembling paddle wheels, which operate at relatively low speed. Forward-curved blade and backward-curved blade fans are usually of the multiblade (10 to 64) type and have a wide range of usefulness.

Axial-flow fans are made in two general types, disk type and propeller type. Disk type fans have plain or curved blades similar to an ordinary household fan. They are usually used for general circulation or exhaust work without ducts. Propeller type fans have blades similar to aeronautical designs.

Centrifugal and Axial Compressors--
Centrifugal compressors or turbo-blowers are widely used to handle large volumes of gas at pressure rises from 0.5 to several hundred lb/sq in. Single casings do not usually contain more than six or seven stages. Two or more casings can be used in series, often with intercoolers. To control the operating range by speed variation, use is made of inlet guide vanes or a blast gate in the discharge line. Close-couple gear mounted compressors are used for relatively clean gases.

The axial compressor has been developed for use with gas turbines. High efficiency and higher capacity are the main advantages of these compressors over centrifugal machines.

Rotary Blowers and Compressors--
Rotary compressors, blowers, and vacuum pumps are machines of the positive-displacement type. Such units are essentially constant-volume machines with variable discharge pressure, depending on the resistance on the discharge side of the system. Rotary compressors are generally classified in several types: straight-lobe, screw, sliding-vane, and liquid-piston.

The straight-lobe units are available for pressure differentials up to about 12 lb/sq in and capacities up to 15,000 cu ft /min. Individual-stage pressure differentials are limited by the shaft deflection, which is kept small to maintain rotor and casing clearance. The screw-type rotary compressor is capable of handling capacities up to 25,000 cu ft /min at pressure ratios of 4:1 and higher. Relatively small diameter male and female rotors whose rotation causes the axial progression of successive sealed cavities allow rotative speeds of several thousand revolutions per minute. The sliding-vane units operate at pressures up to 125 lb/sq in and in capacities up to 2000 cu ft /min. Pressure ratios are generally limited to 4:1. The liquid-piston types are offered as single-stage units for pressure differentials up to about 75 lb/sq in and in capacities up to 4000 cu ft/min in the smaller sizes. These units are widely used as vacuum pumps in wet-vacuum service.

Reciprocating Compressors--
Reciprocating compressors are still the type most widely used in the chemical industry and are furnished either single-stage or multistage. The compressors most commonly used for compressing gases have a crosshead to which the connecting rod and piston rod are connected providing a straight-line motion for the piston rod. Either single-acting or double-acting pistons may be used. Single-acting air-cooled and water-cooled are available in sizes up to about 100 hp and in one, two, three, or four stages for pressures as high as 3500 lb/sq in. In some machine double-acting pistons are used in the first stages and single-acting in the later stages. On multistage machines, intercoolers are provided between stages.

The piston on nonlubricated cylinders is equipped with piston rings and pads of graphitic carbon or Teflon. Plastic packing of a type that requires no lubricant is used on the stuffing box. Special connecting pieces can also be furnished between the cylinder and frame.

The use of high-pressure compressors with discharge pressures of 5000 to 25,000 lb/sq in is increasing. These compressors usually have five, six, seven, or eight stages and the cylinders are steel forgings with single-acting plungers and piston/rod packing of the segmental-ring metallic type. Metallic diaphragm compressors are available for small quantities (up to 10 cu ft/min) for compression ratios as high as 10:1 per stage. These compressors are actuated hydraulically by a plunger-pump.

Ejectors--
An ejector is a simplified type of vacuum pump consisting essentially of a steam nozzle that discharges a high-velocity jet across a suction chamber connected to the equipment to be evacuated. The gas is entrained by the stream and carried into a venturi-shaped diffuser which converts the velocity energy of the steam into pressure energy. Booster ejectors are available with multiple nozzles. Liquid-cooled condensers of either the direct-contact type or surface type usually are used between stages of multistage units to condense the operating vapor from the preceding stage.

Vacuum Systems--
Vacuum systems have been discussed in other sections above with the exception of the diffusion pump. A liquid of low absolute pressure is boiled in a

reservoir within a vessel, ejected within the vessel at high velocity in a downward direction through multiple jets, and condensed on the walls of the vessel which are cooled. Molecules of the gas being pumped are driven downward by collisions with the vapor molecules. The gas molecules are removed through the discharge line by a backing pump such as a rotary oil-sealed unit. By providing a cold trap between the diffusion pump and the region being evacuated, pressures as low as 10^{-7} mm Hg absolute are achieved.

Process Plant Piping

Pipe and Tube--

Pipe and tube are divided into two main classes -- seamless and welded. Seamless pipe refers to pipe made by forging a solid round material, piercing it by simultaneously rotating and forcing it over a piercer point, and further reducing it by rolling and drawing. Seamless pipe and tube are also produced by extrusion, casting into static or centrifugal molds, and by forging and boring.

Welded pipe is made from rolled strips formed into cylinders and seam-welded by various methods. Larger diameters and lower ratios of wall thickness to diameter can be obtained in welded pipe than in seamless pipe (other than cast pipe).

It should be noted that piping systems can be constructed of ferrous metal, non-ferrous metal and non-metallic materials or can be lined with various elastomeric or non-metallic materials.

Valves--

Valve bodies may be cast, forged, machined from bar stock, or fabricated from welded plate. Valves serve to regulate the flow of fluids and to isolate piping without interrupting other connected units.

Gate valves are designed in two types. The wedge-shaped gate, inclined-seat type is most commonly used. The wedge gate is usually solid but may be flexible or split. In the double-disk, parallel-seat type, an inclined-plane device mounted between the disks converts stem force to axial force, pressing the disks against the seats. Gate valves may have non-rising stems, inside-screw rising stems, or outside-screw rising stems. Gate valves are used to minimize pressure drop in the open position and to stop the flow of fluid rather than to regulate it.

Globe valves are designed as either inside-screw rising stem or outside-screw rising stem. The disks are free to rotate on the stems; this prevents galling between the disk and the seat. Guides above and/or below the disk are used to prevent misalignment and spinning. Pressure drop through globe valves is much greater than that for gate valves, and globe valves in horizontal lines prevent complete drainage.

Angle valves are similar to globe valves; the same bonnet, stem, and disk are used for both. They combine an elbow fitting and a globe valve into one unit with a substantial saving in pressure drop.

Diaphragm valves are limited to pressures of about 50 lb/sq in. Fabric-reinforced diaphragms may be made from natural rubber, synthetic rubber, or from rubber faced with Teflon. Plastic bodies, which have low moduli of elasticity compared with metals, are practical in diaphragm valves since alignment and distortion are minor problems. These valves are excellent for fluid containing suspended solids and can be installed in any position.

Size and shape of the port divide plug cocks into the following types: short venturi, reduced rectangular port, long venturi, reduced rectangular port, full rectangular port, and full round port. Plug cocks are limited to temperatures below 500°F. In lever-sealed plug cocks, tapered plugs are raised by turning one lever, rotated by another lever, and reseated by the first lever. In plug cocks other than lever-sealed, the contact area between plug and body is large, and gearing is usually used in sizes 6 in and larger to minimize operating effort.

In ball valves, since the sealing element is a ball, its alignment with the axis of the stem is not essential to tight shutoff. In free-ball valves the ball is free to move axially. In fixed-ball valves, the ball rotates on stem extensions, with the bearings sealed with o-rings. Ball valves in which the ball and seat are inserted from above are known as top-entry ball valves. The others are known as split-body valves.

Butterfly valves occupy less space in the line than any other valves. Relatively tight sealing, without excessive operating torque and seat wear, is accomplished by a variety of methods, such as resilient seats, piston rings on the disk, and inclining the stem. Because fluid pressure distribution tends to close the valve, the smaller manually-operated valves have a latching device on the handle, and the larger ones use worm gearing on the stem. Pressure drop is quite high compared with that of gate valves.

Check valves are used to prevent reversal of flow. Unlike most other valves, check valves are most likely to leak at low pressure than at high pressure. For this reason elastomers are often mounted on the disk. Lift-check valves are made in three styles: vertical, globe, and angle. Tilting-disk check valves are often used for rapid closure at the instant of reversal of flow.

Storage of Liquids

Atmospheric Tanks --

An atmospheric tank is designed to be used within plus or minus a few pounds per square feet of atmospheric pressure. It may be either open to the atmosphere or enclosed. Open tanks may be used to store materials that will not be harmed by water, weather, or atmospheric pollution. Otherwise, a roof, either fixed or floating, is required. Fixed roofs are usually either domed or coned, the latter with intermediate supports. Fixed-roof atmospheric tanks require vents to prevent pressure changes that would otherwise occur from temperature changes and withdrawal or addition of liquid. An effective way to prevent vent loss is to use one of the many types of variable-volume tanks. These are built under American Petroleum Institute (API) Standard No. 650 on Welded Steel Tanks for Oil Storage. They may have floating roofs of the double-deck or single-deck type. There are lifter-roof types in which the roof either has a skirt moving up and down in an annular liquid seal or is connected to the tank shell by a flexible membrane. Floating roofs must have a seal between the roof and the tank shell. Drains must be provided for removal of water and the tank shell must have a "wind girder" to avoid distortion.

Pressure Tanks--
Vertical cylindrical tanks constructed with domed or coned roofs, which operate at pressures above a few pounds per square foot but are still relatively close to atmospheric pressure, can be built according to API Standard 650. The pressure force acting against the roof is transmitted to the shell, which either has sufficient weight to resist it or is reinforced by an anchor ring or heavy foundation. As the size or the pressure goes up, curvature on all surfaces becomes necessary. Shapes used are spheres, ellipsoids, toroidal structures, and circular cylinders with torispherical, ellipsoidal, or hemispherical heads. Tanks in this category, up to a pressure of 15 lb/sq in, can be built according to API Standard 620. The ASME Pressure Vessel Code also contains design specifications.

Pond and Underground Storage--
Low-cost liquid materials, if they will not be damaged by rain or atmospheric pollution, may be stored in ponds. A pond may be excavated or formed by damming a ravine. The soil may require treatment to make it sufficiently impervious or by lining the pond with concrete, plastic film, or other barrier.

Underground storage can be provided by porous media between impervious rock. Cavities can be formed in salt domes and beds by dissolving the salt and pumping it out. Underground chambers are also constructed in frozen earth. Underground tunnel or tank storage is often the most practical way of storing hazardous materials.

Storage of Gases--
Gas is sometimes stored in expandable gas holders of either the liquid-seal or dry-seal type. The liquid-seal holder has a cylindrical container, closed at the top, and varies its volume by moving it up and down in an annular water-filled seal tank. The seal tank may be staged in several lifts. The dry-seal holder has a rigid top attached to the sidewalls by a flexible fabric diaphragm which permits it to move up and down.

Certain gases will dissolve readily in liquids. In some cases where the quantities are not large, this may be practical. The end use requirement of anhydrous or liquid state will predetermine whether this method can be used.

Pressure vessels and pipes can be used for storing gases under pressure. A storage pressure vessel is usually a permanent installation. Storing a gas under pressure not only reduces its volume but also in many cases liquefies it at ambient temperature. Pipes, although not ordinarily considered as storage devices, have been buried in a series of connected parallel lines and used for storage. This avoids the necessity of providing foundations and the earth protects the pipe from temperature extremes. Storage is also obtained by increasing the pressure in operating pipe lines and thus using the pipe volume as a tank.

Low-temperature and cryogenic storage is used for gases that liquefy under pressure at atmospheric temperature. In cryogenic storage the gas is at, or near to, atmospheric pressure and remains liquid because of low temperature. The liquefied gas must be maintained at or below its boiling point. Refrigeration can be used, but the usual practice is to cool by evaporation. At very low temperatures, the tank may have double walls with the interspace evacuated or filled with an insulating material.

HANDLING OF BULK AND PACKAGED SOLIDS

Conveying of Bulk Solids

Screw Conveyors--
The screw conveyor is one of the oldest and most versatile conveyor types. It consists of a helicoid or sectional flight, mounted on a pipe or shaft and turning in a trough. Power to convey is transmitted through the pipe or shaft and is limited by the allowable size of this member. Screw conveyor capacities can approach up to 10,000 cu ft/hr. Typical feed arrangements consist of plain spouts or chutes, rotary cutoff valves, rotary-vane feeders, rack-and-pinion or bin gates, and side inlet gates.

Belt Conveyors--
The belt conveyor is almost universal in application. It can travel for miles at speeds up to 1000 ft/min and handle up to 5000 tons/hour. Belt-conveyor slopes are generally in the 18° to 20° range. Bulk density, temperature, and chemical activity of the conveyed material as well as operating conditions play important roles in belt selection and belt-conveyor design. Belt-conveyor idler and plate-support arrangements can consist of a flat belt on flat-belt idlers or continuous plate or a troughed belt on 20° or 45° idlers or continuous plate.

Bucket Elevators--
Bucket elevators are the simplest and most dependable units for making vertical lifts. They are available in a wide range of capacities and may operate entirely in the open or totally enclosed.

Spaced-bucket centrifugal-discharge elevators are the most common. Mounted on a belt or chain, the buckets are spaced to prevent interference in loading or discharging. This type of elevator will handle almost any free-flowing, fine or small lump material. Spaced-bucket positive-discharge elevators are similar to centrifugal-discharge units except that the buckets are mounted on two strands of chain and are snubbed back under the head sprocket to invert them for positive discharge. These units are designed especially for materials that are sticky or tend to pack. Continuous-bucket elevators are generally used for larger lump materials or materials too difficult to handle with centrifugal-discharge units. Buckets are closed spaced, with the back of the preceeding bucket serving as a discharge chute for the bucket that is dumping as it rounds the head pulley. Supercapacity continuous-bucket elevators are designed for high lifts and large-lump material. They handle high tonnages and are usually operated at an incline to improve loading and discharge operations.

The buckets are generally back-mounted to chain or belt at close intervals. V-bucket elevator-conveyors and skip hoists are also used.

Vibrating or Oscillating Conveyors--
Most vibrating conveyors are essentially directional-throw units which consist of a spring-supported horizontal pan vibrated by a direct-connected eccentric arm, rotating eccentric weights, an electromagnet, or pneumatic or hydraulic cylinder. The purpose of the motion is to throw the material upward and forward so that it will travel along the conveyor path in a series of short hops.

Vibrating conveyors can be classified based on drive characteristics: direct or positive, semipositive, and non-positive drive. All these types transmit vibration to their supporting structures, but semipositive and non-positive drive types reduce vibration effects because thrust is transmitted over the entire support length rather than at a specific point. Semipositive-drive units consist of mechanical and pneumatic/ hydraulic vibrating conveyors while non-positive drive includes mechanical and electrical vibrating conveyors.

Continuous-Flow Conveyors--
The continuous-flow conveyor is a totally enclosed unit that has a relatively high capacity per unit of cross-sectional area and can follow an irregular path in a single plane. These conveyors employ a chain-supported conveying element which means that the material feeding into the conveyor must fall past the chain element and travel in a reverse direction before passing into the actual conveying leg. The closed-belt conveyor, with zipper-like teeth which mesh to form a closed tube, is particularly adaptable to handling fragile materials. Since the belt is wrapped snugly around the material, it moves with the belt and is not subject to any form of internal movement except at feed and discharge. Flight conveyors are generally open designs for rough conveying operations and are available in an almost infinite variety. Apron conveyors are probably the most common chain conveyors. Their main application is the feeding of material at controlled rates, with lump sizes that are large enough to minimize dribble.

Pneumatic Conveyors--
This type of conveyor moves material suspended on a stream of air, over horizontal and vertical distance ranging from a few to several hundred feet. Materials ranging from fine powders through 1/4-in pellets and with densities of 1 to more than 200 lb/cu ft can be processed. Pneumatic conveyors are generally classified according to five basic types: pressure, vacuum, combination pressure and vacuum, fluidizing, and blow tank.

Storage of Solids in Bulk

Storage Bins, Silos, and Hoppers--
A bin is the upper section of a storage vessel and has vertical sides. The hopper, which has at least one sloping side, is the section between the bin and outlet of the vessel. Mass-flow bins feature the most sought-after characteristics of a storage vessel, i.e., unassisted flow whenever the bottom gate is opened. Mass flow means that all the material in the vessel

moves whenever any is withdrawn. A funnel flow-bin may or may not flow but probably can be made to flow. Funnel flow means that only a portion of the material flows when any material is withdrawn. Gates are used to control flow from bins, hoppers, and processing equipment. The rack-and-pinion gate operates manually with a minimum of effort and is easily adapted to electric pneumatic or hydraulic operation. The lever-operated quadrant gate is often used where a quick-opening gate is desired.

Flow-Assisting Devices, Feeders--
Vibrating hoppers are one of the most important and versatile of flow assisters. They are used to enlarge the storage bin opening and to cause flow by braking up material bridges. Two basic types of vibrating hoppers are common: the gyrating kind and the whirlpool type. Screw feeders use a variable-pitch screw to produce a uniform draw of material across the entire hopper opening. Other feeders used to give uniform feed from a bin are belt or apron, table, vibratory, and star. Other methods of aiding bin unloading are rotating arm units and air fluidizing pads.

Packaging of Solids

Bags--
Multiwall paper bags are the most common packages for almost any palleted or powdered material as well as briquettes or bats of solids and are made from plies of kraft paper or from combinations of kraft and special-purpose papers and plastics. Two bag designs are common: the valve and the open-mouth types. The valve kind has both ends closed during fabrication, the filling being done through a small opening in one corner of the bag. The open-mouth has one end closed at the factory, the other being sealed after filling. Open-mouth bags are closed mainly by sewing, although adhesive is applied to the pinch type. Valve bags have the advantage over the open-mouth type in that highly productive filling machines are available for their use.

Valve-Bag Filling Equipment--
Although paper and plastic bags can be filled by a wide variety of equipment, the simultaneous fill-and-weigh (gross weigher) kind predominates. The most important in this category is the fluidizing type. Fluidizing bag fillers meet production requirements ranging from pilot-plant scale through heavy-duty, conveyorized, high tonnage installations. A column of product over a chamber provided with an air pad at the bottom, adjacent to a filling spout, is what causes flow. Fluidizing and pressurizing air is best provided by a positive-displacement blower. Auger or screw-type bag fillers are usually applied to tuck-in sleeve-type valve bags. Single-screw, filling-spout designs with simultaneous fill-and-weigh features are most common. The centrifugal belt-type packer is used extensively on granular or pelleted products whose bulk densities range from 25 to 100 lb/cu ft.

Impeller-type fillers are used extensively for finely divided materials. These fillers contain an impeller that turns in a casing to move the product into the bag. Most impeller machines are installed with conveyors. Gravity-type fillers find occasional use in marginal operations where investment must be limited and performance is not critical. Only free-flowing pellets and granules in the 1/4-in to 30-mesh range can be handled practically.

Open-Mouth-Bag-Filling Equipment--
With open-mouth bag-filling equipment, preweigh scales discharge to a chute system to which a bag is attached. The kinetic energy of the charge as it reaches the bottom permits the bag to stand without lateral support on a closing-machine conveyor. The filled bag is then dropped to a short-belt conveyor that passes the bag through a closing machine. Conventional multiwall, paper, open-mouth bags are closed by sewing, the pinch bottom type by hot-melt adhesive. Drum and bulk-box filling consists of three operations: setting up, filling and weighing, and closing.

Weighing and filling can be done manually or automatically. The most common installation consists of a conveyor line with a platform scale at a central location. Pouches, small bags, and cartons can also be filled. Also, small packages can be handled at a high rate via the form-fill-seal type of packaging. This method involves two main functions: a weigh cycle and a package make-fill cycle. At present, form-fill-seal is limited to products having reasonably free-flowing particles, with low dust concentrations.

SIZE REDUCTION/ENLARGEMENT

Crushing and Grinding Equipment
A wide variety of size-reduction equipment is available. A practical classification of crushing and grinding equipment is given in Table A-1.

Jaw Crushers--
These may be divided into three main groups: the Blake, with movable jaw pivoted at the top, giving greatest movement to the smallest lumps; the Dodge, with the movable jaw pivoted at the bottom, giving greatest movement to the largest lumps; and the overhead eccentric, with the swing jaw mounted directly on an eccentric shaft so that it receives a downward as well as a forward motion.

Gyratory Crushers--
The gyratory crusher, used in most large hard-ore and mineral crushing applications, consists of a cone-shaped pestle oscillating within a larger cone-shaped mortar or bowl. Primary crushers have a steep cone angle and a small reduction ratio. Secondary crushers have a wider cone angle and spreads the wear over a wider area. The three general types of gyratory crusher are the suspended-spindle, the supported-spindle, and the fixed-spindle types.

Roll Crushers--
In a roll crusher, two rolls of the same diameter are rotated toward each other at the same or different speeds. One of the shafts moves in fixed bearings, the other in movable bearings. The distance between the rolls is adjustable. A number of single-roll crushers are also available.

Toothed shredders are used for friable or plastic materials. The action is largely tearing rather than compressing. The sawtooth crusher has two shafts geared together at differential speeds. Each shaft carries sawtooth and spacer assemblies. The size of the product can be controlled by the spacing of the saws and the peripheral speeds. The horizontal rotary crusher has a

Table A-1. Types of Size-Reduction Equipment

A. Jaw crushers:
   1. Blake
   2. Overhead eccentric
   3. Dodge
B. Gyratory crushers:
   1. Primary
   3. Cone
C. Heavy-duty impact mills:
   1. Rotor breakers
   3. Cage impactors
D. Roll crushers:
   1. Smooth rolls (double)
   2. Toothed rolls (single and double)
E. Dry pans and chaser mills
F. Shredders:
   1. Toothed shredders
   2. Cage disintegrators
   3. Disk mills
G. Rotary cutters and dicers
H. Media mills:
   1. Ball, pebble, rod and compartment mills:
      a. Batch
      b. Continuous
   2. Autogenous tumbling mills
   3. Stirred ball and sand mills
   4. Vibratory mills
I. Medium peripheral-speed mills:
   1. Ring-roll and bowl mills
   2. Roll mills, cereal type
   3. Roll mills, paint and rubber types
   4. Buhrstones
J. High-peripheral-speed mills:
   2. Pin mills
   3. Colloid mills
K. Fluid-energy superfine mills:
   1. Centrifugal jet
   2. Opposed jet
   3. Jet with anvil

toothed cone supported on a horizontal shaft for preliminary crushing; final crushing takes place between close-fitting sections at the base of the cone.

Impact Breakers--
Impact breakers include heavy-duty hammer crushers and rotor impact breakers. With the hammer crusher, pivoted hammers are mounted on a horizontal shaft, and crushing takes place by impact between the hammers and breaker plates. The continuous movement of the traveling breaker plate in a non-clog hammer mill forces the feed into the crushing path of the hammers. In the rotor impactor, the rotor is a cylinder to which is affixed a tough steel bar. Free impact breaking occurs against this bar or on rebound from the walls of the device. Dual-rotor impact breakers, reversible impactors, vertical-impact pulverizers, ring-type granulators, and cage mills have also been built to handle certain types of rock or to produce a desired product.

Pan Crushers--
The pan crusher consists of one or more grinding wheels or mullers revolving in a pan; either the pan or the mullers may remain stationary with the other(s) driven. Iron scrapers or plows at a proper angle feed the material under the mullers which are made of tough alloys. Where a solid pan bottom is used in place of a perforated screen bottom and the ground material is discharged through a slot in the rim, the machine is described as a rim-discharge grinder. The Chambers dry pan uses air cylinders to regulate the grinding pressure under each of the muller tires.

Tumbling Mills--
Ball, pebble, rod, tube, and compartment mills have a cylindrical or conical shell, rotating on a horizontal axis, and are charged with a grinding medium, such as balls of steel, flint, porcelain, or with steel rods. As the cylindrical shell is rotated horizontally about its axis, size reduction is effected by the tumbling of the grinding medium on the material between them. The ballmill differs from the tube mill by being short in length. The tube mill is usually long in comparison with its diameter, uses smaller balls, and produces a finer product. The compartment mill consists of a cylinder divided into two or more sections by perforated partitions; preliminary grinding takes place at one end and finish grinding at the discharge end. Rod mills deliver a more uniform and more granular product than other revolving mills. The pebble mill is a tube mill with flint or ceramic pebbles as the grinding medium and may be lined with non-metallic liners. Multicompartmental mills are also available which feature grinding of coarse feed to finished product in a single operation, wet or dry. Autogenous tumbling mills employ coarse lump feed as the grinding medium while it is itself being ground.

Mill-feeders used to pass the feed into the mill without backspill are of several types. A feed chute is generally used for dry grinding, this consisting of an inclined chute sealed at the outer edge trunnion and down which the material slides to pass through the trunnion and into the mill. A screw feeder may also be used when dry grinding, consisting of a short section of screw conveyor which extends part way into the opening in the feed trunnion and conveys the material into the mill. For wet grinding, several different types of feeders are available; the scoop feeder attached to and rotating with the mill trunnion and which dips into a stationary box to pick up the material and pass it into the mill; a drum feeder attached to and rotating with the feed trunnion, having a central opening into which the material is fed, and an internal deflector or lifter to pass the material through the trunnion into the mill; or a combination drum and scoop feeder, where the new feed to the mill is fed through the central opening of the drum while the scoop picks up the oversize being returned from a classifier to a scoop box well below the center line of the mill.

Non-Rotary Ball or Bead Mills--
In the stirred type, a central paddle wheel or impeller armature stirs the media at speeds from 100 to 1500 rpm. In the vibratory type, an eccentric motion is imparted either to an armature or to the shell at speeds up to 1800 rpm while the media oscillate in one or more planes. Vibratory mills may grind dry, but most stirred mills are restricted to wet milling. Among stirred mills can be found the Sweco dispersion mill, the Attritor and the Bureau of Mines Mill. Vibratory mills include the Vibro-Energy and Podmore-Boulton mills and the Vibratom.

Particle-Size Classifiers Used with Grinding Mills--
Ball mills or tube mills can be operated in closed circuit with external air classifiers with or without air sweeping being employed. Many types of grinders are air-swept and are so closely coupled with their classifiers that the latter are termed internal classifiers. Dry classifiers or screens are used primarily in crusher circuits, since they are most effective down to 4-mesh. Most dry-milling circuits use air classifiers. Rotating blades are the main elements of several types of classifiers. They set up a centrifugal motion that tends to throw coarser particles outward. Closed-circuit wet classifiers are generally used in large-scale operations.

Ring-Roller Mills---
Ring-roller mills, also referred to as ring-roll mills or medium-speed mills, are equipped with rollers that operate in conjunction with grinding rings. Grinding takes place between the surfaces of the ring and rollers. Either the ring or rollers may be stationary. These mills are built with and without internal screen classification, in which the material is discharged from the grinding chamber through a surrounding screen. In bowl mills, the rollers do not touch the ring. The raw material from the feeder drops on the bowl where, owing to centrifugal force of rotation, it is forced to the periphery between the ring and stationary rollers, where it is pulverized.

Hammer Mills--
Hammer mills are operated at high speeds. The rotor shaft carries hammers that can be T-shaped elements, stirrups, bars or rings fixed or pivoted to the shaft or to disks fixed to the shaft. The rotor runs in a housing containing grinding plates or liners. A cylindrical screen or grating, serving as an internal classifier, usually encloses all or part of the rotor. The grinding action results from impact and attrition between lumps or particles of the material being ground, the housing, and the grinding elements. A large number of variations of the hammer mill are available: the Mikri-Pulverizer, Aero Pulverizer, Disintegrator, Tornado mill, Fitz mill, Turbo-Pulverizer, Turbo mill, pin mill, Kolloplex mill, Entoleter

impact mill, Imp Pulverizer, Automatic pulverizer, Hurricane pulverizer-classifier, Mikro-Atomizer, Mikro-Bud pulverizer, Mikro-ACM pulverizer, and the Pulvocron.

Disk Attrition Mills--
The disk or attrition mill uses steel disks incorporating interchangeable metal or abrasive grinding plates rotating at high speeds. Grinding takes place between the plates which may operate in a vertical or horizontal plane. The assembly, comprising a shaft, disk, and grinding plate, is called a runner. Feed material enters a chute near the axis, passes between the grinding plates, and is discharged at the periphery of the disks. A form of disk attrition mill is the double-disk mill. In general, single-runner mills are used for the same purposes as double-runner mills except that they will accept a coarser feed stock and their range of reduction is more limited. Buhrstone mills are attrition mills with hard circular stones serving as grinding media and are still employed for special cereals and grains.

Dispersion and Colloid Mills--
This is a special class of mills employed for dispersion and colloidal operations. They operate on the principle of high-speed fluid shear. Their value lies in ensuring a breakdown of agglomerates or, in the case of emulsions, the shearing of fluid phases to produce dispersed droplets of fine size, around 1 micron. Colloid mills fall into four main groups: the hammer or turbine, the smooth-surface disk, the rough-surface type, and value or orifice devices. Examples are the Morehouse mill, Premier mill, Charlotte mill, Gaulin colloid mill, and Manton-Gaulin mill.

Fluid Energy or Jet Mills--
Fluid-energy mills may be classified in terms of the nature of the mill action. In one class of mills, the fluid energy is admitted in fine high-velocity streams at an angle around a portion or all of the periphery of a grinding and classifying chamber. In this class are the Micronizer, Jet Pulverizer, Reductionizer, and Jet-o-Mizer. In the other class the fluid streams convey the particles at high velocity into a chamber where two streams impact upon each other. The Majac-jet pulverizer and other mills are in this category. In either class, there is a high energy release as the particles travel around the periphery of the grinding-classifying chamber. This high order of turbulence causes the particles to grind upon themselves and to be ruptured.

Size Enlargement Equipment

Pressure Compacting Equipment--
Compacting is carried out in two classes of equipment. There are confined pressure devices (molding, tableting, roll presses) in which internal motion and shear of the particles are incidental to their consolidation in closed molds or between two surfaces; and extrusion devices (pellet mills, screw extruders) in which material undergoes definite shear and mixing as it is consolidated while being pressed through a die.

Agglomeration Equipment--
Powders with the correct amount of a liquid binder can be formed into regular agglomerates by tumbling, vibrating, shaking, or paddle mixing. Rotating drums and disks are the equipment most commonly used industrially. An inclined pan or disk agglomerator consists basically of an inclined rotating disk equipped with a rim to contain the agglomerating charge. Under centrifugal action, the agglomerating material in a pan travels in spiral rings of decreasing diameter until balls of the required size discharge over the lip. A drum agglomerator consists of an inclined rotary cylinder powered by a variable-speed drive. The pitch of the drum is sufficient to produce movement of material down the length of the drum. As with the inclined pan, the inside surface of the drum must be rough to ensure proper tumbling action. Also, scraper or cutter bars run longitudinally down the inside of the cylinder. A traveling grate is commonly used to bond by treating at high temperatures in conjunction with other agglomeration processes.

HEAT GENERATION AND TRANSPORT

Fired Process Equipment
In direct-fired combustion equipment, the flame and/or products of combustion are used to achieve the desired result by direct contact with another material. Common examples are rotary kilns, open-hearth furnaces, and submerged-combustion evaporators. In indirect-fired combustion equipment, the flame and products of combustion are separated from any contact with the principal material in the process by means of metallic or refractory walls. Examples are vaporizers, heat exchangers, and melting pots. Other types of combustion equipment do not fall into either classifications given above.

Direct-Fired Process Equipment--
Gaseous direct-fired heaters are often used where the products of combustion do not seriously affect the process stream. Liquid heaters of the direct-fired type are used under the same conditions as those of the gaseous type but are more common since the combustion products do not absorb to a large extent and may not be objectionable. Solid heaters are common in the metallurgical industries. Radiant heat transfer is often used effectively either by porous-wall refractory burners or by multiple-wall burners. Convection heat transfer can be improved by direct flame impingement. Chemical reactors, such as oxidation reactors are direct-fired process equipment used to achieve a desired chemical reaction. Partial-combustion reactors utilize a less-than-stoichiometric supply of oxidizer to produce a desired chemical product. This process usually involves pyrolysis.

Indirect-Fired Process Equipment--
Metallic indirect-fired heaters utilize the heat-transfer medium of a thin metal wall to transfer thermal energy from the combustion process to a liquid, gas, or encapsulated solid. Non-metallic heaters utilize refractories or silica glass as the heat-transfer medium. Steam generators are divided into two categories; industrial and power utility steam boilers. Industrial steam boilers include fire-tube boilers and water-tube boilers. Fire-tube boilers are characterized by the containment of the products of combustion within the tubes of the boiler. Water-tube boilers contain the vaporizing water within the tubes while the combustion products surround the tubes. Utility boilers used in central-station power plants commonly operate in the

supercritical range. They are usually of water-tube design and may include forced circulation, reheat, divided furnaces, once-through flow, and pressurized furnaces. Vaporizers are used to achieve high temperatures without the high pressures associated with steam and are frequently organic fuel fired. Reformers are externally heated tubes in which endothermic chemical reactions are carried out.

Incinerators--
Incinerators consist of a refractory or air-cooled chamber into which solid waste can be feed for burning. Atmospheric quality restrictions require control of emissions in the form of bag houses and such. Large incinerators are usually charged mechanically; the waste is moved through the furnace on grates so that combustion conditions are controlled and made more nearly uniform than with the smaller batch-fired units. Other types of large incinerators employ rotating kilns, rocking grates, inclined reciprocating grates, and drum grates to provide agitation of the burning refuse.

Heat Transport

Thermal-Liquid Process System--
The Dowthern process systems are either gravity return or pumped return. The most desirable, since no moving parts are required, is the gravity system, in which the vapor rises from the vaporizer to the heated vessels, condenses, and flows back to the vaporizer by gravity. Various control and safety features are recommended, such as three-way valves, special tungsten-steel-alloy relief valve springs, the so-called Hartford loop and a storage tank for containment of the entire system charge. Mineral oil systems using aromatic mineral oils must always be in a closed system with a cold-oil expansion tank to prevent hot oil from contacting air. Conventional mineral oils are not affected by contact with air. The heater may be direct-fired with combustion gases passing over the tubes through which the heated oil circulates, or electrically heated with the oil flowing through narrow channels over the heating source. All hot-oil systems use forced circulation usually by centrifugal pump. Therminol fluid systems use either of two basic heater designs: liquid-tube type or fire-tube type. In the former, the Therminol is pumped through the tubes at a definite flow rate as it is heated. In the latter, the fluid flows through the shell. For large flow rates, a centrifugal circulating pump is used. The expansion tank should accommodate the heated, expanded fluid.

Heat Pipes--
The structural elements of a heat pipe are a closed outer vessel, a porous capillary wick, and a working fluid. The basic phenomena of evaporation, condensation, and surface tension pumping in a capillary wick permit the heat pipe to transfer latent heat continuously without the help of external work. Glass or metal tubes may be used for the gastight containment vessel. Demonstrated working fluids include water, acetone, glycerin, ammonia, Freon, molten salts, and molten metals.

HEAT TRANSFER EQUIPMENT

Shell-and-Tube Heat Exchangers

Shell-and-tube-type exchangers constitute the bulk of the unfired heat-transfer equipment in chemical process plants. Important features of the principal types of exchangers are summarized in Table A-2.

Heat transfer equipment can be designated by type (e.g., fixed tube sheet, outside packed head, etc.) or by function (chiller, condenser, cooler, etc.). Almost any type of unit can be used to perform any or all of the functions listed in Table A-3. Other types and variations of the shell-and-tube heat exchanger are the bent-tube fixed-tube-sheet, bayonet-tube, spiral-tube, falling-film, and Teflon heat exchangers. The tube bundle is the most important part of a tubular heat exchanger. The tubes generally constitute the most expensive component of the exchanger and are the one most likely to corrode. Tube sheets, baffles or support plates, tie rods, and usually spacers complete the bundle. Baffles are provided for heat-transfer purposes and can be of several types: segmental, orifice, disk & doughnut, window-cut, impingement, and longitudinal.

Table A-2. Features of Principal Shell-and-tube-type Exchangers

Type of design	Fixed tube sheet	U-tube	Packed lantern-ring floating head	Internal floating head (split backing ring)	Outside-packed floating head	Pull-through floating head
Provision for differential expansion	Expansion joint in shell	Individual tubes free to expand	Floating head	Floating head	Floating head	Floating head
Removable bundle	No	Yes	Yes	Yes	Yes	Yes
Replacement bundle possible	No	Yes	Yes	Yes	Yes	Yes
Individual tubes replaceable	Yes	Only those in outside row	Yes	Yes	Yes	Yes
Tube cleaning by chemicals inside and outside	Yes	Yes	Yes	Yes	Yes	Yes
Interior tube cleaning mechanically	Yes	Special tools required	Yes	Yes	Yes	Yes
Exterior tube cleaning mechanically:						
Triangular pitch	No	No	No	No	No	No
Square pitch	No	Yes	Yes	Yes	Yes	Yes
Hydraulic-jet cleaning:						
Tube interior	Yes	Special tools	Yes	Yes	Yes	Yes
Tube exterior	No	Yes	Yes	Yes	Yes	Yes
Double tube sheet feasible	Yes	Yes	No	No	Yes	No
Number of tube passes	No practical limitations	Any even number posible	Limited to one or two passes	No practical limitations	No practical limitations	No practical limitations
Internal gaskets eliminated	Yes	Yes	Yes	No	Yes	No

Table A-3. Functions of Heat Transfer Equipment

Equipment	Function
Chiller	Cools a fluid to a temperature below that obtainable if water only were used as a coolant. It uses a refrigerant such as ammonia or Freon.
Condenser	Condenses a vapor or mixture of vapors, either alone or in the presence of a non-condensable gas.
Partial condenser	Condenses vapors at a point high enough to provide a temperature difference sufficient to preheat a cold stream of process fluid. This saves heat and eliminates the need for providing a separate preheater (using flame or steam).
Final condenser	Condenses the vapors to a final storage temperature of approximately 100°F. It uses water cooling, which means the transferred heat is lost to the process.
Cooler	Cools liquids or gases by means of water.
Exchanger	Performs a double function: (1) heats a cold fluid by (2) using a hot fluid which it cools. None of the transferred heat is lost.
Heater	Imparts sensible heat to a liquid or a gas by means of condensing steam or Dowtherm.

Table A-3. (Continued)

Equipment	Function
Reboiler	Connected to the bottom of a fractionating tower, it provides the reboil heat necessary for distillation. The heating medium may be either steam or a hot process fluid.
Thermosiphon reboiler	Natural circulation of the boiling medium is obtained by maintaining sufficient liquid head to provide for circulation
Forced-circulation reboiler	A pump is used to force liquid through the reboiler.
Steam generator	Generates steam for use elsewhere in the plant by using the available high-level heat in tar or a heavy oil.
Superheater	Heats a vapor above the saturation temperature.
Vaporizer	A heater which vaporizes part of the liquid.
Waste-heat boiler	Produces steam; similar to steam generator, except that the heating medium is a hot gas or liquid produced in a chemical reaction.

Other Heat Exchangers

Double-Pipe Exchangers--
These are well adapted to high-temperature, high-pressure applications because of their relatively small diameters. Double-pipe exchangers can be made by inserting one pipe within another and then welding the outer jacket to the inner pipe. Double-pipe sections permit true countercurrent flow, a particular advantage when very close temperature approaches are required. Multitube double-pipe sections are also available. Scraped-surface exchangers have a rotating element with spring-loaded scraper blades to scrape the inside surface. Generally a double-pipe construction is used with the scraper inside the inner pipe, where the process fluid flows, and the cooling/heating medium in the outer pipe.

Plate-Type Exchangers--
Plate-type exchangers are available in several different forms: spiral, plate, brazed-plate fin, and plate fin-and-tube types. The spiral-plate exchanger is made from a pair of plates rolled to provide two relatively long rectangular passages for fluids in countercurrent flow. The spiral design is compact and the continuous path eliminates flow reversals, by-passing, and differential-expansion problems. The plate-type heat exchanger consists of standard plates, which serve as the heat-transfer surfaces, and a frame to support them. Pressure drop is low and interleakage of fluids is impossible. The brazed-plate-fin heat exchanger is made up of a stack of layers, the heat-transfer surfaces, with each layer consisting of a corrugated fin between flat metal sheets sealed off on two sides by channels or bars to form one passage for the flow of fluid. The plate fin-and-tube type utilizes rectangular fins that are pierced, formed, belled, and stacked before tubes are inserted into the fin collars and expanded to produce the plate fin-and-tube surface. No solder or brazing metal is used.

Air-Cooled Heat Exchangers--
In the usual design, ambient air is forced or induced by a fan to flow across a bank of externally finned tubes. A typical air cooler has a horizontal section containing finned tubes, a steel supporting structure with plenum chambers and fan ring, axial-flow fan, drive assembly, and miscellaneous accessories. The 1-in outside diameter tube is most commonly used while bimetallic tubing, also often used, consists of an inner tube of material required by the process corrosive conditions and an aluminum outer tube from which most of the metal has been cold-extruded into high fins. The tube bundle consists of headers, finned tubes, structural-steel side channels, and supports. Air-cooled overhead condensers have been designed and installed above distillation columns as integral parts of distillation systems. Cascade coolers consist of a series of tubes, mounted horizontally, one above the other. Cooling water from a distributing trough drips over each tube and then flows to a drain.

Heating/Cooling of Tanks--
Pipe tank coils are commonly shop-fabricated in helical and spiral coils while a hairpin pattern is generally field-fabricated. The helical coils are used principally in process tanks and pressure vessels. Fin-tube coils are used for fluids which have poor heat-transfer characteristics. Teflon

immersion coils are available with braided or unbraided 0.10-in tubes. Braiding increases overall heat-transfer efficiency. Bayonet heaters have bayonet-tube elements consisting of an outer and an inner tube. These elements are inserted into tanks and process vessels for heating or cooling. External coils and tracers are used to heat or cool tanks, vessels, and pipe lines externally. Jacketing is often used for vessels needing frequent cleaning.

Evaporators

By far the largest number of industrial evaporators employ tubular heating surfaces. Circulation of liquid past the heating surface may be induced by boiling or by mechanical means. Evaporators can also be arranged as single-effect or multiple-effect. Single-effect evaporators are used where the required capacity is small, steam is cheap, the material is significantly corrosive, or the vapor is so contaminated that it cannot be reused. Single-effect evaporators may be operated batch, semibatch, continuous-batch, or continuously. Multiple-effect evaporation is used for economizing on energy consumption. Most such evaporators operate on a continuous basis, although a continuous-batch cycle may be employed. The feed to a multiple-effect evaporator can be by backward-feed, forward-feed, mixed-feed, or parallel-feed operation.

Forced-Circulation Evaporators--
The use of a pump to ensure circulation past the heating surface makes it possible to separate the functions of heat transfer, vapor-liquid separation, and crystallization. The pump withdraws liquor from the flash chamber and forces it through the heating element back to the flash chamber. Circulation is maintained regardless of the evaporation rate. Highest transfer is obtained when the liquid is allowed to boil in the tubes. By far the largest number of forced-circulation evaporators are of the submerged-tube type.

Short-Tube Vertical Evaporators--
The body of this type evaporator is a vertical cylinder and the tubes are expanded into horizontal tube sheets that span the body diameter. Circulation past the heating surface is induced by boiling in the tubes, which are usually 2 to 3 in in diameter and 4 to 6 ft long. The circulation rate through the tubes is many times the feed rate and the circulation and heat transfer are strongly affected by the liquid level.

Long-Tube Vertical Evaporators--
This evaporator consists of a simple one-pass vertical shell-and-tube heat exchanger discharging into a relatively small vapor head. Feed enters at the bottom of the tube, and starts to boil part way up the tube. The mixture of liquid and vapor leaving the top at a high velocity impinges against a deflector placed above the tube sheet. This deflector is effective both as a primary separator and as a foam breaker. The falling-film version of this evaporator eliminates problems of hydrostatic head although feed distribution to the tubes is a problem.

Horizontal-Tube Evaporators--
In this type of evaporator, the steam is inside and the liquid outside the

tubes. Low entrainment loss is the primary aim and small shell diameters allow operation at high steam and vapor pressures. Usually descaling is done semiautomatically with ease.

Miscellaneous Forms of Heating Surfaces--
Jacketed kettles, frequently with agitators, are used when the product is very viscous, the batches are small, intimate mixing is required, and/or ease of cleaning is important. The channel-switching evaporator alternates the duty of either side of the heating surface periodically from boiling liquid to condensing vapor so that scale formed can be dissolved. Agitated thin-film evaporators employ a heating surface consisting of one large-diameter tube that may be either straight or tapered, horizontal or vertical. Liquid is spread on the wall by a rotating assembly of blades.

Evaporators Without Heating Surfaces--
The submerged-combustion evaporator makes use of combustion gases bubbling through the liquid as the means of heat transfer. It consists simply of a tank to hold the liquid, a burner and gas distributor that can be lowered into the liquid, and a combustion-control system. Disk or cascade evaporators consist of a horizontal shaft on which are mounted disks perpendicular to the shaft or bars parallel to the shaft. The assembly is partially immersed in the liquor so that films of liquor are carried onto the hot-gas stream as the shaft rotates. Flash evaporators also require no heating surfaces. An example is a recrystallizing process for separating salts having normal solubility curves from salts having inverse solubility curves.

Evaporator Accessories--
The vapor from the last effect of an evaporator is usually removed by a condenser. Surface condensers are employed where mixing of condensate with condenser cooling water is not desired. For the most part they are shell-and-tube condensers with vapor on the shell side and multi-pass flow of cooling water on the tube side. The most common type of direct-contact condenser is the countercurrent barometric condenser, in which vapor is condensed by rising against a rim of cooling water. Another type of direct-contact condenser is the jet or wet condenser, which makes use of high-velocity jets of water to both condense the vapor and force the non-condensable gases out the discharge pipe. Vent systems are also required to dispose of non-condensable gases present in the evaporator as a result of leakage, air dissolved in the feed, or decomposition reactions in the feed.

Heat Exchangers for Solids

This equipment is so constructed that the solids load (burden) is separated from the heat-carrier medium by a wall, i.e., heat transfer is by the indirect mode. The two phases are never in direct contact. Some of the devices handle the solids burden in a static or laminar-flowing bed. Other devices resemble continuously agitated kettles with respect to their heat transfer operation.

Equipment for Solidification--
A frequent operation in the chemical field is the removal of heat from a material in a molten state to effect its conversion to the solid state. When performed batchwise, it is termed casting, but when done continuously, it is termed flacking. The table type heat exchanger is a simple flat metal sheet with slightly upturned edges and jacketed on the underside for coolant flow It is still widely used in casting operations. The agitated-pan type is a circular flat surface with jacketing on the underside for coolant flow and with a means of stirring to sweep over the heat-transfer surface. It is a batch-operation device. The vibration type is used when the burden has special needs. When solidification has been completed and vibrators started, the intense vibratory action of the deck structure breaks free the friable cake, shatters it, and conveys the lumps up over the dam to discharge. The metal-belt type (the "water-bed" conveyor) features a thin wall, a well-agitated fluid side for a thin water film, a conveyor belt "floated" on the water with the aid of guides, no removal knife, and cleanability. The rotating-drum type is available in the single- and twin- or double-drum configuration. The double-drum is best suited to thin (1/4 inch) cake production. The rotating-shelf type is best suited for thick-cake production and burdens having an indefinite solidification temperature.

Equipment for Fusion of Solids--
The thermal duty here is the opposite of solidification. The horizontal-tank type is used to transfer heat for melting dry powdered solids or drying divided solids. The vertical agitated-kettle type is used to melt to the liquid state and provide reaction heat for solids that vary greatly in "body" during the process. The mill type is a power device as well as a device for indirect heat transfer, employed to knead and heat a mixture of dry powdered ingredients with the objective of reacting and reforming the mixture by fusion to a consolidated product.

Heat-Transfer Equipment for Sheeted Solids--
Cylinder heat-transfer units, sometimes called can dryers or drying rolls, are distinguished from drum dryers in that they are used for solids in flexible continuous-sheet form. A cylinder dryer may consist of one large cylindrical drum; more often it comprises a number of drums arranged so that a continuous sheet of material passes over them in series. Cylinder dryers usually operate at atmospheric pressure and by steam heat.

Heat-Transfer Equipment for Divided Solids--
Most equipment for this service is adapted from material handling systems. Some types of equipment are convertible from heat removal to heat supply by simply changing the temperature level of the fluid or air. Others require auxiliary or structural changes. The fluidized-bed type, known as the cylindrical fluidizer, operates with a bed of fluidized solids. Heating applications are many and varied. The moving-bed type uses a single-pass tube bundle in a vertical shell with the divided solids flowing by gravity in the tubes. It is used in specialized applications. The agitated-pan type is used in a batch operation mode and has found wide application. Both heating and cooling are feasible but the greatest use has been for drying. Kneading devices are closely related to the agitated pan but differ as being primarily

mixing devices with heat transfer a secondary consideration. Heat transfer is provided by jacketed construction of the main body. Shelf devices are used for freeze drying by sublimation. Rotating-shell devices are installed horizontally and are found in several modifications: plain, flighted, and tubed-shell. These devices are used for cooling, heating and drying, and are the workhorses for heat processing divided solids in the large capacity range. The metal-belt type of conveyor-belt devices has been adapted for indirect heat-transfer service with divided solids. Spiral conveyor devices have been adapted for evaporation duty in a batch-operating mode. Several types are available: the jacketed solid-flight, small-spiral-large-shaft, and the large-spiral, hollow-flight type. The double-cone blending device provides mixing action and very effective burden exposure to the heat-transfer surface, which is provided by a jacket on the shell. Vibrator-conveyor devices are suited for heating and cooling and are available in the following types: heavy-duty jacketed, coolant-spraying type, light-duty jacketed, and Mix-R-Step. Elevator devices, pneumatic conveying devices, and vacuum-shelf units are also employed for divided solids heat-transfer operations.

EVAPORATIVE COOLING AND REFRIGERATION

Evaporative Cooling

Evaporative cooling involves the exposure of water surface to air in varying degrees. The heat-transfer process consists of latent heat transfer owing to vaporization of a small portion of the water and sensible heat transfer owing to the difference in temperature of water and air.

Mechanical-Draft Towers--

Two types of mechanical-draft towers are in use today--the forced draft and the induced draft. In the forced-draft tower the fan is mounted at the base and air is forced in at the bottom and discharged at low velocity through the top. The induced-draft tower is the most common type. It is further classified into counterflow or cross-flow design, depending on the relative flow directions of water and air. The counterflow arrangement is more efficient since the coldest water contacts the coldest air thus obtaining maximum enthalpy potential. The cross-flow tower increases air flow primarily by lengthening the tower to increase the air-flow cross-sectional area. Performance of a given cooling tower is governed by the ratio of the weights of air to water and the time of contact between the water and air. Time of contact is governed largely by the time required for the water to discharge from the nozzles and fall through the tower to the basin.

Atmospheric Cooling Towers--

An atmospheric cooling tower is one in which water cooling is obtained primarily by natural wind movement through the structure. The cooling capacity of any tower, with a given wet-bulb temperature and wind velocity, varies with the water concentration.

Natural-Draft Towers--

Natural draft or hyperbolic-type towers are primarily suited to very large cooling-water quantities. Air flow is due largely to the difference in density between the cool inlet air and the warm exit air. The air leaving the stack is lighter than the ambient air and a draft is created by chimney effect thus eliminating the need for mechanical fans.

Cooling Ponds--
Where large ground areas are available, cooling ponds may be constructed within earthen embankments up to 10 feet high. The soil should be impervious or provided with soil sealant layers. Location in a flat area is desirable. Four principal heat transfer processes are involved: evaporation, convection, pond radiation, and solar radiation. The area of pond required for a given cooling load is almost independent of pond depth. Factors considered to affect pond performance are air temperature, relative humidity, wind speed, and solar radiation.

Spray Ponds--
Spray ponds provide an arrangement for lowering the temperature of water by evaporative cooling and, thereby, greatly reduce the cooling area required in comparison with a cooling pond. A spray pond uses a number of nozzles which spray water into contact with the surrounding air. The pond should be placed distance should be provided from the outer nozzles to keep spray from being carried over the sides of the basin.

Air Conditioning

Air conditioning is the process of treating air to control simultaneously its temperature, humidity, cleanliness, and distribution to meet the requirements of the conditioned space. Basically, an air-conditioning system consists of a fan unit which forces a mixture of fresh outdoor air and room air through a series of devices which act upon the air to clean it, increase or decrease its temperature, and increase or decrease its water-vapor content or humidity. Modern air-conditioning equipment generally falls into three classes: (1) self-contained, (2) unit-type, and (3) central system.

Self-Contained Equipment--
Self-contained equipment, also called unitary or packaged equipment, may be further classified as room air conditioners, residential units, and commercial or industrial units. These units contain a complete cooling system in one package, consisting of a compressor, condenser, evaporator, fan, filter, and controls. A heating coil and humidifier may also be included. The units may have either air-cooled or water-cooled refrigerant condensers. Cooling water may be either once-through or recirculated; in the latter case a cooling tower or spray pond is used to remove heat from the water.

Unit-Type Equipment--
The unit-type air conditioners are designed for one room only and require a remotely connected refrigeration machine or other means of cooling, and a remotely connected source of heat. These units consist of one or more motor-driven centrifugal fans, a cleanable or disposable air filter, a water or direct-expansion coil, outside air damper, and controls.

Central-Station Systems--
These units may serve one or several areas with conditioned air being

supplied to the various areas through ductwork. Either water or direct-expansion refrigerant coils or air washers may be used for cooling. Steam or hot-water coils are available for heating, and humidification may be provided by target-type water nozzles, pan humidifiers, steam humidifiers, air washers, or sprayed coils.

Heat Pumps--
Year-round heating and cooling is provided by the heat pump. The evaporator of a standard refrigeration system removes heat from the supply air to the conditioned space during the cooling season and dispels it through the condenser to outdoor air or to water. During the heating season the cycle is reversed. The heat is then "pumped" to the condenser which provides the heat to the air supply to the conditioned space. Any air conditioning-refrigeration system can be converted into a heat pump.

Refrigeration

Mechanical refrigeration is the process of lowering the temperature of a substance below that of its surroundings. The chemical process industry is a major user of refrigeration facilities. Refrigeration is used to remove heat of chemical reactions, to liquefy process gases, for gas separation by distillation and condensation, and to purify products by preferential freeze-out of one component from a liquid mixture.

Vapor Compression Systems--
The single-stage refrigeration cycle is used for single-stage vapor compression. The four basic components in the system are the compressor, condenser, expansion valve, and evaporator. As liquid refrigerant flows through the evaporator, heat is absorbed from a fluid being cooled and the refrigerant boils. The low-pressure vapor is compressed. The pressure and temperature levels are increased to a point where the superheated vapor can be condensed by the cooling media available. The liquid refrigerant flows from the condenser to an expansion valve where its pressure and temperature are reduced to those in the evaporator. The cycle is thus completed. The multi-state refrigeration cycle avoids high compression ratios. Multi-stage systems are of two basic types--compound and cascade. The compound cycle using a reciprocating compressor would employ a booster or first-stage compressor and a gas-liquid interstage cooler in a basic refrigeration cycle. Only one refrigerant is used in the compound cycle. A compound centrifugal system also uses series flow of a single refrigerant. The main difference is that a flash-type desuperheater is used instead of the gas-liquid interstage cooler. The cascade cycle involves two separate refrigeration systems which are interconnected in such a manner that one provides a means of heat rejection for the other. Cascade systems permit the use of different refrigerants in each cycle of the cascade to produce low temperatures. A cascade condenser serves as the condenser of the low stage and the evaporator for the high stage. The cascade condenser is usually sized so that the evaporating temperature for the high stage is 10°F below the condensing temperature of the low stage. Compound and cascade cycles can be advantageously combined using either centrifugal or reciprocating compressors.

Vapor Compression Equipment--
Two main types of compressors used for basic refrigeration vapor cycles utilizing the common refrigerants are: the dynamic-type centrifugal compressors, and the positive-displacement-type reciprocating compressor. Large-capacity systems are most economically handled by centrifugal machines. Reciprocating compressors can be applied to systems of 150 tons or less (air-conditioning requirements) and for specialized low-temperature work when inlet gas volumes are not too large. Centrifugal compressors use impeller wheels to impart energy to the gas being compressed. Impellers should have a minimum gas flow of approximately 1500 to 2000 cu ft/min. The overall head against which the compressor must operate is determined by the suction and condensing properties of the refrigerant selected. The capacity is controlled by impeller speed, inlet guide vanes, and suction dampers. Reciprocating compressors are primarily of the vertical, single-acting arrangement. These compressors are classified into two basic categories: open type, where the compressor is driven externally through a crankshaft extending outside the crankcase, and hermetic type whereby the compressor and motor are connected by a common shaft, and both are located internally within the compressor housing. The open type may be either directdriven or belt-driven while the hermetic type eliminates the need for a mechanical shaft seal. The majority of reciprocating compressors are of the high-speed, multicylinder design. A steam jet may be used to replace the compressor in refrigeration systems. The refrigerant is then water vapor, and the steam jet removes it from the flash tank (evaporator), compresses it, and delivers it to the condenser.

Absorption Systems--
Present-day systems of this type mainly use water as the refrigerant and lithium bromide as the absorbent. This absorption process utilizes two basic factors to produce the refrigeration effect: (1) water will boil and flash cool itself at low temperatures when it is maintained at a high vacuum, and (2) certain salt solutions, such as that of lithium bromide, are hygroscopic and will absorb water vapor. An absorption machine consists of five main components: evaporator, absorber, solution heat exchanger, generator, and condenser. Three circuits are involved in the absorption unit operation: water as a refrigerant is pumped to the evaporator; lithium bromide as the absorbent is circulated over the absorber tubes, through the heat exchanger, and to the generator; cooling water flows in series initially through the absorber tubes and partially through the condenser tubes.

DISTILLATION COLUMNS

The separation process called distillation utilizes various kinds of devices called plates or trays to bring the two phases, vapor and liquid, into intimate contact. The trays are stacked one above the other and enclosed in a cylindrical shell to form a column. The feed material which is to be separated into fractions is introduced at one or more points along the column shell. Because of the difference in gravity between the vapor and liquid phases, the liquid runs down the column, cascading from tray to tray, while the vapor goes up the column contacting the liquid at each tray.

The liquid reaching the bottom of the column is partially vaporized in a heated reboiler to provide reboil vapor which is sent back up the column. The remainder of the bottom liquid is withdrawn as the bottom product. The vapor reaching the top of the column is cooled and condensed to a liquid in the overhead condenser. Part of this liquid is returned to the column as reflux to provide liquid overflow. The remainder of the overhead stream is withdrawn as the overhead or distillate product. The lighter (lower-boiling) components tend to concentrate in the vapor phase while the heavier (higher-boiling) components tend toward the liquid phase.

Azeotropic Distillation

An azeotrope is a liquid mixture which exhibits a maximum or a minimum boiling point relative to the boiling points of surrounding mixture compositions. Azeotropic distillation refers to those processes where a component is added to form an azeotrope with one or more of the feed components and that azeotrope is removed as either the overhead or bottom product. The design is somewhat more complex than that for simple distillation because of the additional design variables resulting from the additional feed; both the solvent/feed ratio and the location of the solvent-entry point must be specified. The azeotropic column would have a reboiler but open steam or a reboiler could be used on the solvent-recovery distillation column.

Extractive Distillation

Extractive distillation refers to those processes where a high-boiling solvent is added to alter the relative volatility of the components in a feed mixture. The design of an extractive-distillation column is usually simpler than for an azeotropic column. To maintain a high concentration of solvent throughout most of the column, the solvent must always be introduced to the column above the fresh-feed stage, almost always within a few trays of the top. The actual concentration of the solvent will change abruptly at the fresh-feed-introduction point if a liquid feed is used. A vapor feed is sometimes used to avoid dilution of the descending solvent. The reflux at the top of the extractive-distillation column also tends to dilute the solvent by increasing the amount of non-solvent material in the liquid overflow. The inherent advantage of higher reflux rates must, in the case of extractive distillation, be balanced against the effect on the solvent concentration and the changes in relative volatilities that occur.

Petroleum Distillation

The principles of multi-component distillation apply to the case of petroleum distillation. The initial separation of crude oil into its various cuts is accomplished by atmospheric distillation in a pipe-still distillation unit. This unit is mainly a rectifying column from which a number of side-stream products are withdrawn as well as overheads and bottoms. Separate steam strippers are often used with each side stream to eliminate "light ends," which are returned to the main column. The bottoms from the initial atmospheric-pressure distillation of the crude may next be subjected to a vacuum-distillation operation to make an overhead product suitable for further refining. To identify various chemical entities present in a given petroleum fraction, three types of laboratory distillations are commonly used. These are the true-boiling-point (TBP) distillation, the American Society for Testing and Materials (ASTM) distillation, and the equilibrium flash vaporization (EFV) distillation. Specialized distillation equipment is specified for each distillation procedure.

Batch Distillation

Batch distillation is the process of separating a specific quantity (the charge) of a liquid mixture into its components. Many large installations feature a batch still when the material to be separated is high in solids content or contains tars and resins that might plug or foul a continuous unit. The still for a simple batch distillation consists of a heated vessel, a condenser, and one or more receiving tanks. Material is charged into the vessel, called the pot, and the liquors are brought to boiling. The vapors are condensed and collected in a receiver. The rate of vaporization is sometimes controlled to prevent "bumping" of the charge and to avoid overloading the condenser, but other controls are minimal. To obtain products with a narrow composition range, batch distillation with rectification is used. Components of a rectifying still are the pot, a rectifying column, a condenser, a means of splitting off a portion of the distillate as reflux, and one or more receivers. The temperature of the distillate is controlled in order to return the reflux near the column temperature to permit a true indication of reflux quantity and to improve the column operation. A subcooling heat exchanger is then used for the remainder of the distillate. The column may also operate at pressure or vacuum, in which case appropriate devices must be included. Batch columns are also used extensively for azeotropic batch distillations. Instead of using multiple columns, cuts are segregated in storage for later reuse or distillation.

Molecular Distillation

Molecular distillation involves a vapor path that is unobstructed and a condenser that is separated from the evaporator by a distance less than the mean free path of the evaporating molecules. A simple molecular still, with an evaporative efficiency between 0.4 to 0.6 can be formed by an electrically heated tray suspended in an evacuated test tube. (The efficiency of transfer of a conventional still is 0.001 to 0.0001.) Heat is applied beneath the still and the ceiling is cooled by ice or a blast of air. Liquid condenses on the ceiling and drops into the receiver. Temperature of distillation and hazard of decomposition are reduced in proportion to the lowered pressure of the molecular still. Commercial apparatus include falling-film stills and centrifugal stills. Falling-film stills have been made in single and multiple units, ranging in size from a few centimeters to 50 cm diameter, 2 to 10 meters high, with throughputs ranging from 1 to 60 liters/hr. The distilland is admitted through a metering device into the vacuum where it is degassed in one or more preliminary vessels and then allowed to pass onto the walls of a heated polished metal cylinder stationed within a concentric cooled condensing cylinder.

The space between the two is maintained under high vacuum (1 to 5 microns) by an oil-filled diffusion pump backed by a mechanical vacuum pump or a multistage steam ejector. A form of the falling-film still uses wipers of carbon or Teflon held in a cylindrical cage which rotates between evaporator and condenser to greatly improve its performance. Centrifugal stills consist of a housing or base plate covered by a lid or dome which serves as the air-cooled condenser. A flat or conical plate on which the distilland will be spread by centrifugal force is supported on a shaft that passes through a bearing and stuffing box attached in the base plate. A radiant electrical heater warms the plate. Distilland is admitted from a preliminary degasser through a feed pipe to a depression at the center of the rotor where it is spun rapidly outward in an exceedingly thin, uniform layer. At the edge, the distilland, spent after evaporation of the volatile constituents, is picked up by a scoop or collected by a concentric gutter. The condensate formed on the cooled dome is collected at the edge of the base plate.

GAS ABSORPTION TOWERS

Gas absorption is the process by which one or more soluble components of a gas mixture are dissolved in a liquid. The absorption may be a purely physical phenomena or may involve solution of the material in the liquid followed by reaction with one or more constituents in the liquid solution. The equipment used for continuous contacting of a vapor and a liquid can be a tower filled with solid packing material, an empty tower into which the liquid is sprayed and through which the gas flows, or a tower that contains a number of bubble-cap, sieve, or valve-type plates. In general, the gas and liquid streams flow countercurrent to each other in order to obtain the greatest concentration driving force and therefore the greatest rate of absorption. Occasionally, absorption operations are carried out in spray columns, wetted-wall columns, stirred vessels, or other types of equipment.

Usually, packed columns are chosen for corrosive materials, for low pressure drop, for pilot-plant or small-scale operations, and for liquids that foam badly. Plate columns are preferred for large-scale operations, for low liquor rates, and where internal cooling is desired. In packed towers, packings found generally useful are 1/2 to 2-inch ceramic or carbon resins, 1-inch saddles, 3-inch spiral or partition rings, drip-point tile, and wood grids. The pressure drop through a packed tower can frequently be less than for a plate tower and still obtain adequate vapor-liquid contact. On the other hand, the total weight of a plate tower is usually less than that of a packed tower designed for the same duty. Cooling coils can readily be installed on plates. Also, plate towers are generally preferred for operations that require a large number of transfer units.

LIQUID EXTRACTION SYSTEMS

Industrial processes most commonly employ continuous countercurrent multistage contact in a series of mixers and settlers or a tray-type tower, or countercurrent differential contact in a continuous tower of the packed or mechanically agitated type. In laboratory practice, concurrent batch operation by single or simple multistage contact is common.

Countercurrent Multistage Contact

Fresh solvent and feed enter at opposite ends of a series of extraction stages. Extract and raffinate layers pass continuously and countercurrently from state to state through the system. Any number of stages may be employed, the more common numbers being three to six. The system may be composed of a series of mixers, each with its separate settler, or some form of tray column may be used.

Continuous Countercurrent Differential Contact

If one of the phases is subdivided and allowed to pass continuously and countercurrently through the other phase, which is not dispersed, continuous countercurrent differential operation is obtained. Either the solvent or the feed may be subdivided and allowed to pass through the other. This may be conducted either in a packed column or in mechanically agitated towers. If equilibrium is approached, this operating system theoretically gives maximum "efficiency."

Countercurrent Extraction with Reflux

The extract composition may be raised beyond that merely in equilibrium with the feed by providing reflux at the used to bring about the full extent of purification required. In cyclic extract end of the extractor. Reflux is supplied by returning to the extractor a portion of the extract layer from which the solvent has been wholly or partly removed. This requires that the feed enter at an intermediate point in the tower or system.

Double-Solvent (Fractional) Extraction

A mixture of components B and C to be separated may be subjected to continuous countercurrent extraction with immiscible solvents A and D. A multistage extractor (of whatever form) is then fed more or less centrally with B and C, frequently dissolved in a small amount of one of the solvents, especially if the substances to be separated are solids. Solvents A and D enter at opposite ends of the extractor, and flow countercurrently, component B leaving preferentially with solvent A and component C with solvent D. Reflux of separated solute may be used at either or both ends of the extractor, which reduces the number of stages required.

Single Contact

The simplest method and that most common on the laboratory scale is to bring each quantity of solvent and feed together in one contact and then recover the product and solvent without further extraction. This method is the least effective and is rarely feasible on an industrial scale. The degree of separation between components of the feed is poor. Operations may be either batch or continuous.

Simple Multistage Contact

In this procedure (sometimes called crosscurrent multistage contact), the total quantity of solvent to be used is divided into several portions. The feed is then treated with each of these portions of fresh solvent in a series of successive steps or stages; raffinate from the first extraction step is treated with fresh solvent in a second stage and so on. However, unless the carrier in the feed is largely immiscible with the extracting solvent, separation between the components of the feed tends to be poor. An exceedingly large proportion of solvent is required to obtain a high degree of extraction by this method, and concentration of solute in the extract layer becomes increasingly dilute in succeeding stages. The method may be operated intermittently with a single mixer-and-settler unit or continuously with a series of such units. The well-known Soxhlet extraction method corresponds to simple multistage contact with an infinite number of stages.

ADSORPTION AND ION EXCHANGE EQUIPMENT

Adsorbents are natural or synthetic materials of microcrystalline structure whose internal pore surfaces are accessible for selective combination of solid surface. This combining effect is identified as physical adsorption. Ion exchange consists of a solid phase containing bound groups that carry an ionic charge, either positive or negative, in conjunction with free ions of opposite charge that can be displaced. Most ion exchangers currently in large-scale use are based on synthetic resins.

Adsorption

Sorption processes are employed in liquid-phase and gas-phase treatments. Fixed-bed contacting with periodic regeneration is generally employed, although countercurrent systems and batch treatment both find use in individual cases. Three major types of solid-fluid sorption equipment are in use: (1) batch units; (2) fixed beds of adsorbent through which the process fluid passes, with periodic interruption for regeneration; and (3) systems which provide for countercurrent or concurrent movement of adsorbent and fluid in a continuous or quasi-continuous (intermittent) operation.

Batch Operations--

It is often advantageous to carry out sorbent-liquid contact in batch equipment. Batch methods are well adapted to laboratory use and have also been applied to larger scale operations. Laboratory methods can involve the use of contact adsorbents that are subsequently separated by filtration. Batch tests are often conducted on portions of adsorbent or ion-exchange material intended for larger-scale use. The contact filtrations of lubricating oils is a process used to remove colored and carbon-forming materials from lubricant stocks. Acid and clay treatment is employed in the process. The particle size of the clay is substantially finer than can be used in the alternative method of percolation treatment in fixed beds. Diatomaceous earth may be used as a filter precoat or may be mixed with the slurry to improve its filtering properties. Mixer-settler operations can be accomplished in the same vessel by using large-particle sorbent material through which a liquid can drain quite readily. With the vessel filled with a charge of process liquor, gentle agitation is usually obtained by sparging the slurry with air. Draining of the liquor from the settled bed is subsequently hastened by use of a reverse gas flow. Where the process liquor is a slurry rather than a clear liquid, slurry-granular sorbent contact is generally preferred to percolation-type contact. After sufficient contact, the process liquor may be centrifuged or filtered. Subsequently, the sorbent is recovered by hydraulic classification or wet screening.

Fixed Beds--
The most frequently used method of fluid-solid contact for sorption operations is in columnar units, with the solid particles closely packed in a relatively fixed arrangement. For regenerating an adsorbent, the regenerant may be a gas instead of a liquid and, if so, the off-gas passes through a condenser where the solvent is recovered. Joint use of adsorbents may be used to bring about the full extent of purification required. In cyclic operations, two fixed-bed adsorber or exchanger units are provided, so that one is on stream while the other is being regenerated. Thus a fixed-bed assembly can simulate a completely countercurrent system by a valve arrangement that provides the proper succession of saturant and regenerant streams to each bed.

Continuous and Quasi-Continuous Operations--
Continuous-flow units involving the transport of solid particles have been proposed for both gas and liquid sorption operations. Carbon-treatment systems, using the typical granular activated-carbon adsorber incorporated in a system designed to permit thermal regeneration of the carbon, may be classified as a continuous, countercurrent unit. The adsorber tank is usually a vertical cylindrical pressure vessel, with fluid distributors at top and bottom. The column is filled with granular carbon. Fluid flow is upward, and carbon is intermittently displaced downward by opening a valve at the bottom and injecting a measured slug of carbon into the top of the vessel. The exhausted slug is transferred to the "sweeten-off" tank, where residual product is displaced. It next is dewatered and fed to the regeneration furnace, where it eventually returns to the adsorber.

Ion Exchange

Ion exchange resins are used in fixed beds, intermittent countercurrent columns, and slurry units. They consist simply of pressure columns equipped with piping, valves, and accessory equipment to permit regeneration of the resins in place. A typical fixed-bed ion exchange column consists of a vertical cylindrical pressure vessel of lined steel or stainless steel. Spargers are provided at the top and bottom and frequently a separate distributor is used for regenerant solution. The resin bed is supported by the screen of the bottom distributor or by a support bed of graded quartz. Externally, the unit is provided with a valve manifold to permit downflow operation, upflow backwashing, injection of regenerant, and rinsing of excess regenerant. Counterflow regeneration can be achieved by regenerating countercurrent to the service flow for better results than same direction regeneration. The typical ion exchanger contains only one ionic type of resin, either cation or anion, and regeneration is normally accomplished with one chemical solution. In a mixed-bed ion exchanger, however, a cation and an anion resin are contained in the same column, usually in approximately equimolar proportion. The unit has a screened distributor at the plane or interface between the two resins so that they may be separately regenerated without removing them from the column. To permit uniform columnar exhaustion of the resin, the tank often has a flat bottom. Mixed-bed deionizers are often used to "polish" condensate. Pumps are used to recirculate condensate through the deionizer. Countercurrent columns are used in ion-exchange equipment in which the resin and the contacting solutions flow countercurrently, but intermittently, to produce quasi-continuous ion exchange. Two competitive designs of continuous ion exchangers are the Higgins design and the Asahi process.

MISCELLANEOUS SEPARATION PROCESSES

Leaching

Leaching is a process of removing a solute or solutes from a solid by the use of liquid solvent. Leaching equipment is principally divided into two major classes: (1) that in which the leaching is accomplished by percolation; and (2) that in which particulate solids are dispersed into a liquid, and subsequently separated from it. Each includes batch and continuous units.

Percolation--
Percolation is carried out in batch tanks and in several designs of continuous extractors. The batch percolator is a large circular or rectangular tank with a false bottom. The solids are dumped into the tank to a uniform depth and are sprayed with solvent until their solute content is satisfactorily reduced and are then excavated. Countercurrent flow of the solvent through a series of tanks is common. Some tanks operate under pressure. A series of pressure tanks operating with countercurrent solvent flow is called a diffusion battery. Continuous percolators consist of moving-bed equipment, including single-deck and multi-deck rake classifiers, bucket-elevator contactors, and horizontal-belt conveyors, examples of which are the Bollman-extractor, the Rotocel extractor, and the Kennedy extractor.

Dispersed-Solids Leaching--
Equipment for leaching fine solids by dispersion and separation includes batch tanks agitated by rotating impellers or by air, and a variety of continuous devices. Batch stirred tanks are agitated by coaxial impellers (turbines, paddles, or propellers) are commonly used for batch dissolution of solids in liquids. The agitator is most efficient if it just gently circulates the solids across the tank bottom. After leaching, the solids may be separated by settling and decantation of the supernatant extract, or by external filters, centrifuges, or thickeners. Ores of certain metals are often batch-leached in large air-agitated vessels known as Pachuca tanks. Continuous dispersed-solids leaching can be achieved by a vertical-plate extractor, exemplified by the Bonotto extractor, which consists of a column divided into cylindrical compartments by equispaced horizontal plates. The solids fed to the top plate, fall to each lower plate in succession while the solvent flows upward through the tower. The solids are discharged by a screw conveyor and compactor. Gravity sedimentation tanks operated as thickeners

can serve as continuous contacting and separating devices in which fine solids may be leached continuously. Also, impeller-agitated tanks can be operated as continuous leaching tanks, singly or in series.

Screw-Conveyor Extractors--
The screw-conveyor extractor is actually neither a percolator nor a dispersed-solids extractor. The Hildebrandt total-immersion extractor uses a helix surface perforated so that solvent can pass through countercurrently. The screws compact the solids during their passage through the unit. A somewhat similar design uses a horizontal screw section for leaching and a second screw in an inclined section for washing, draining, and discharging the extracted solids.

Crystallization

Crystallization may be carried out from a vapor, melt, or solution. Most industrial applications of crystallization involve crystallization from solutions. Crystallizers in this presentation are classified according to the means of suspending the growing product.

Mixed-Suspension, Mixed-Product-Removal Crystallizer--
This type of equipment, sometimes called the circulating-magma crystallizer, is by far the most important in use today. In a crystallizer, sizing is normally done on the basis of the volume required for crystallization or of special features required for the proper product size. Although surface-cooled crystallizers of this type are available, most users prefer crystallizers employing vaporization of solvents or of refrigerants. A forced-circulation evaporative crystallizer utilizes a tube-and-shell heat exchanger and a circulating pipe through which a slurry leaving the main body is pumped. The liquor, heated in the exchanger, is returned to the body and mixed with the body slurry, raising its temperature and causing boiling at the liquid surface. The supersaturation created causes deposits on the circulating pipe. The feed is admitted into the circulating line after withdrawal of the slurry. A similar type, the conispherical magma crystallizer, employs a radial rather than tangential recirculation inlet. A number of designs have been developed that use circulators located within the body of the crystallizer, thereby reducing the head against which the circulator must pump. An example is the draft-tube-baffle evaporator-crystallizer. For some materials, it is possible to employ a surface-cooled crystallizer which uses a forced-circulation tube-and-shell exchanger in direct combination with a draft-tube crystallizer body. Proper selection of temperature controls, circulating pump flows, and internal baffles are critical. For some applications, it is necessary to go to such low temperatures that a direct-contact-refrigeration crystallizer is used. The refrigerant is admixed with the slurry being cooled in the crystallizer so that the heat of vaporization of the refrigerant cools the slurry by direct contact.

Mixed-Suspension, Classified-Product-Removal Crystallizer--
Classification of the product in this system is normally done by means of an elutriation leg suspended beneath the crystallizing body. Introduction of clarified mother liquor to the lower portion of the leg fluidizes the

particles prior to discharge and selectively returns the finest crystals to the body. Although such a feature can be included on many types of classified-suspension or mixed-suspension machines, it is most commonly used with the forced-circulation evaporative crystallizer and the draft-tube-baffle crystallizer.

Classified-Suspension Crystallizer--
This equipment is also known as the growth or Oslo crystallizer and is characterized by the production of supersaturation in a circulating stream of liquor. Supersaturation is developed in one part of the system by evaporative cooling or by cooling in a heat exchanger and it is relieved by passing the liquor through a fluidized bed of crystals. The fluidized bed may be contained in a simple tank or in a more sophisticated vessel arranged for a pronounced classification of the crystal sizes.

Scraped-Surface Crystallizer--
A number of crystallizer designs have been developed employing direct heat exchange between the slurry and a jacket or double wall containing a cooling medium. The heat-transfer surface is scraped or agitated in such a way that deposits cannot build up. An example of such a device is the Swenson-Walker crystallizer consisting of a trough 24 in wide with a semicircular bottom and a cooling jacket welded to the outside. Within the crystallizer is a spiral agitator turning 3 to 10 rpm, with a blade/wall clearance of 1/16 in. The solution is fed into one end continuously and crystal slurry overflows from a concentrated solution to a dilute solution. Dialyzers are generally from the opposite end. The double-pipe scraped-surface crystallizer consists of a double-pipe heat exchanger with an internal agitator fitted with spring-loaded scrapers that wipe the wall of the inner pipe. The cooling liquid passes between the pipes. This equipment can be operated in continuous or recirculating batch manner.

Tank Crystallizers--
Static-tank crystallizers employ an unagitated tank for accepting the hot feed solution that is allowed to cool either by natural convection and radiation or by surface cooling through coils in the tank or a jacket on the outside of the tank. Sometimes rods or wires are suspended in the tank to provide centers for crystallization. The agitated-tank crystallizer uses a propeller or turbine and a cooling system to greatly increase the capacity of a tank crystallizer. Agitation sufficient to suspend the crystals introduces the possibility of increased nucleation. Although vigorous agitation reduces the tendency for crystal buildup on the cooling coils, it does not eliminate it. However, the application of Teflon-tube heat exchangers to cooling-type tank crystallizers appears promising.

Sublimation

Sublimation is the vaporization of a substance from the solid into the vapor state without formation of an intermediate liquid phase. For sublimation to occur, the vapor pressure of the subliming components must be greater than their partial pressures in the gas phase in contact with the solid. Sublimation is accomplished by heating the solid or controlling the gaseous environment in contact with the solid or both. The environment can be controlled either by vacuum operation or by using a nonreactive diluent to lower the partial pressures of the subliming components. The former operation is called vacuum while the latter is known as carrier or entrainer sublimation. Vacuum sublimation is inherently a batch operation, whereas entrainer sublimation can be and usually is conducted as a continuous process. A characteristic of sublimation apparatus is that vaporization and condensation of the sublimable components must be achieved. Typical equipment which has served as the sublimer for simple sublimation is jacketed pan dryers, vacuum rotary dryers, vacuum shelf dryers, direct-fired retorts, and Herreshoff roasting furnaces. Solids-condensing equipment often consists of field-fabricated tanks equipped with mechanical scrapers, brushes, or vibrators for removing the condensed solids. Among devices suitable for continuous-fractional sublimation is one that provides reflux by use of circulating inert solid particles upon which the sublimable components are deposited as a thin film and another that obtains reflux by mechanical transportation or free fall of the sublimable solids.

Adsorptive-Bubble Separation Methods

These methods are based on the selective adsorption or attachment of material on the surfaces of gas bubbles passing through a solution or suspension. In most of the methods, the bubbles rise to form a foam or froth which carries the material off overhead. Adsorptive bubble separation methods consist of foam separation, including foam fractionation and flotation, and nonforming adsorptive bubble separation, including solvent sublation and bubble fractionation. Foam fractionation usually implies the removal of dissolved material while flotation implies the removal of solid particulate material. In solvent sublation an immiscible liquid is placed atop the main liquid to trap the material deposited by the bubbles as they exit. In bubble fractionation the ascending bubbles deposit their adsorbed or attached material at the top of the pool as they exit.

Both batch and continuous-flow operation of adsorptive-bubble separation systems are available. Four alternative modes of continuous-flow operation with a foam-fractionation column are: (1) the simple mode, (2) enriched operation, (3) stripping operation, and (4) combined operation. The equipment consists basically of a column containing the feed material, the resultant foam and a foam overflow leading to a foam breaker. Inlet lines to the column include those for the pool feed and gas, which passes through a simple round perforations and the plate material can be corrugated to bubbler or sparger, and an outlet line for bottoms. The foam breaker contains exit lines for gas and foamate (collapsed foam). Collapse of the overflowing foam can be accomplished by chemical means, by a rotating disk or perforated basket, by discharging onto Teflon instead of glass, or by sound or ultrasound.

Membrane Processes

A membrane is a thin barrier separating two fluids. The barrier prevents hydrodynamic flow so that gas or liquid transport through the polymeric membrane is by sorption and diffusion. The driving force for transport is either pressure or concentration. There are quite a variety of designs of membrane separation units and they consist of tube bundles, stacks, bi-flow stacks and spiral modules.

Dialyzers--
Dialysis is the transfer of solute molecules across a membrane by diffusion designed for continuous operation. There are two principal types of commercial dialyzers: the tank type and the filter-press type. The Cerini dialyzer is an example of the tank type. It consists of impregnated-membrane bags made of mercerized cotton hanging in a tank of liquor, with solvent circulating on the inside of the bags. The filter-press type is a newer model with a considerably greater efficiency than that of the tank type. In this type of dialyzer, vertical membranes are sandwiched in between alternate liquor and solvent frames, the liquor and solvent being fed to the bottom and top of these frames, respectively. Dialyzate and diffusate are removed through channels located at the top and bottom.

Reverse Osmosis Devices--
Reverse osmosis, or ultrafiltration, separates a solute from a solution by forcing the solvent to flow through a membrane by applying a pressure greater than the normal osmotic pressure. There are four common designs of membrane modules. The plate-and-frame unit consists of thin plastic plates covered on both sides by membrane which are sealed around the edge to prevent leaks. The flat surfaces of the plates contain small grooves through which the permeate flows after passing through the membrane. The permeate eventually flows into a central tube at the stack and is collected through this tube. The spiral-wound system consists of an envelope of membrane around a matrix of glass beads held together by a plastic resin. The matrix is connected to a perforated tube at one end of the membrane envelope. The membrane envelope-matrix construction is then wound around the tube in a jelly-roll fashion, and the assembly is inserted into a canister-type pressure vessel. Feed solution flows over the membrane and the purified solvent passing through the membrane flows to a collection system via the inner tube. Tubular devices consist of parallel bundles of rigid-walled, porous or perforated tubes. The inside walls are lined with the membrane. Pressurized feed flows inside the tubes and ultrafiltrate drips off the outside surfaces and is collected in troughs or vessels. The hollow-fiber membrane module consists of fibers with an outside diameter of 25 to 250 microns and a wall thickness of 5 to 50 microns. Input water, under high pressure, flows over the outside surface of the fibers. The preheated water then flows out through the base of the fibers and is collected as product.

LIQUID-GAS SYSTEMS

Process equipment utilized for liquid-gas contacting is based on a combination of operating principles of three categories: (1) countercurrent, co-current, or crossflow mode of flow, (2) differential or integral gross mechanism of transfer, and (3) gas or liquid continuous phase.

Liquid-Gas Contacting Systems

Plate Columns--
Plate columns utilized for liquid-gas contacting may be classified according

to mode of flow in their internal contacting devices: (1) crossflow plates and (2) counterflow plates. The crossflow plate utilizes a liquid downcomer and is more generally used than the counterflow plate. Common liquid-flow patterns with crossflow plates include standard crossflow, reverse flow, double pass, double pass-cascade, and four pass. Downcomers are used to control liquid-flow pattern. Most new designs of crossflow plates employ perforations for dispersing gas into liquid on the plate. These perforated plates are called sieve plates or value plates. The most common gas disperser, the bubble cap, is being displaced by simple or valve-type perforations. In counterflow plates, liquid and gas utilize the same openings for flow. Thus, there are no downcomers. Openings are usually partially segregate the gas and liquid flow. Types of such plates used commercially are perforated, slotted, and perforated-corrugated. A counterflow plate often used for contacting gases with liquids containing solids is the baffle plate or shower deck. Typically the plate is half-moon in shape and is sloped slightly in the direction of liquid flow. Gas contacts the liquid as it showers from the plate.

Packed Columns--
Packed columns for gas-liquid contacting are used extensively for absorption operations and, to a limited extent, for distillations. The packing material may fill the column in a random orientation or in a carefully positioned fashion. The packed column is characteristically operated with counterflow of the phases. A typical column consists of a cylindrical shell containing a support plate for the packing material and a liquid-distributing device designed to provide effective irrigation of the packing. Devices may be added to the packed bed to provide redistribution of liquid that might channel down the wall. Several beds may be used in the same column shell. The following are typical packings used in commercial packed columns: Raschig ring, Lessing ring, Berl saddle, Intalox saddle, Tellerette, and Pall ring. Initial distribution of liquid at the top of the packed bed is accomplished by a device that spreads the liquid uniformly across the top of the packing. Several types of liquid distributors are available: the perforated pipe, trough-type, orifice, and weir-riser.

Liquid-Dispersed Contactors--
Spray devices are the most common type of liquid-dispersed gas liquid contactors. A second type is the baffle-plate or shower-deck column. Spray systems operate with a high degree of back mixing of the phases. Three types of spray systems are in common use: the spray column, the cyclonic spray, and the venturi scrubber.

Phase Dispersion

Liquid-in-Gas Dispersions-
There are three diverse liquid-in-gas dispersions that are of great practical interest in the process industries: (1) sprays produced by nozzles and other atomizing systems for combustion, mass/heat transfer, or coating of surfaces; (2) entrainment generated by gas bubbling through a liquid as in a distillation tower; and (3) fogs generated from gases that are supersaturated. Most industrial spray nozzles fall into three categories: pressure nozzles, two-fluid nozzles, and rotary devices. Characteristic

spray nozzles are: whirl-chamber hollow cone, solid cone; oval-orifice fan, deflector jet, impinging jet, by-pass, poppet, two-fluid, and vane/rotating disk.

Gas-in-Liquid Dispersion--
The dispersion of gas as bubbles in a liquid or in a plastic mass is effected for one of the following purposes: (1) gas-liquid contacting, (2) agitation of the liquid phase, or (3) foam or froth production. Gas-liquid contacting is usually accomplished with conventional columns or with spray absorbers, but for systems containing solids likely to plug columns, strongly exothermic absorptions, or treatments involving a readily soluble gas, gas dispersers may be used to advantage. Agitation by a stream of gas bubbles rising through a liquid is employed in tanks of large volume or unsymmetrical shape. Foam production is important in froth-flotation separations. A number of devices and methods are used to produce gas-in-liquid dispersions. Spargers employ the simple method of dispersing gas in a liquid contained in a tank by introducing the gas through an open-end standpipe, a horizontal perforated pipe, or a perforated plate at the bottom of the tank. Porous septa, that is, porous plates, tubes, disks, or other shapes made by bonding or sintering together carefully sized particles of carbon, ceramic, polymer, or metal are frequently used for gas dispersion, particularly in foam fractionators. Precipitation of a gas from a supersaturated solution generally results in a fine dispersion of bubbles throughout the liquid. Generation of fine, well-dispersed bubbles will occur if a dissolved or finely divided suspended material is decomposed to yield a gas. Fluid attrition systems, such as nozzles and pipeline contactors, develop turbulence during the rapid flow of fluid through the devices and thus can be used to disperse a gas in the flowing liquid. Cascade systems use a stream of liquid-falling through a gas into a pool to entrain, under proper conditions, a volume of gas and disperse it into the pool. Mechanical agitators, such as the wire-whip agitator, the Turbo-gas-absorber, the Permaerator bio-oxidation aerator, and the turbine- and propeller-type surface aerators, are used very effectively to disperse gas into liquid or to create foam.

Phase Separation

Liquid-in-Gas Systems--
The mechanisms which may be used for separating mist and spray particles from gases are: (1) gravity settling, (2) centrifugal force, (3) inertial impaction, (4) flow-line interception, (5) diffusional deposition, (6) electrostatic precipitation, (7) thermal precipitation, and (8) sonic agglomeration. Many types of collection equipment use more than one type of collection mechanism; so equipment is usually classified by type rather than mechanism. Gravity settlers are among the simplest types of equipment for separating particles from gases. They consist of large vessels in which the velocity of the gas stream is reduced. Cyclone separators use centrifugal force to separate entrained liquid particles from gases. Typical centrifugal separators are: (1) Multiclone, (2) Cutaway Thermix ceramic tube, (3) Van Tongeren cyclone, (4) Sirocco type D collector, and (5) the horizontal

steam separator. Impingement separators entrain liquid particles by placing an obstruction in the gas stream to divert the flow of gas and using the particles' forward momentum relative to the gas flow for impingement and collection. Typical impingement separators are: (1) the jet impactor, (2) wave plate, (3) staggered channels, (4) "vane-type" mist extractor, (5) Peerless line separator, (6) Strong separator, (7) Karbate line separator, (8) type E horizontal separator, (9) PL separator, and (10) the wire-mesh demister (Reference: Chemical Engineers' Handbook, Fifth Edition). Typical separators using impingement in addition to centrifugal force are: (1) the Hi-eF purifier, (2) Flick separator, (3) Aerodyne tube, (4) Aerodyne collector, and (5) the type RA line separator. Packed-bed separators are usually not available as standard commercial units but are designed for specific applications. Coke boxes are an example of this type of separator. Fiber mist eliminators use a packed bed of fibers retained between two concentric screens. Mist particles are collected on the surface of the fibers and become a part of the liquid film which wets the fiber. Scrubbers are devices in which a liquid is employed to assist in the removal employed. Electrostatic precipitators for mist are generally of the wire-in-tube type rather than the plate-type used for dust. The ionizing wires consist of lead-covered steel cables suspended from regular metal bars. Means are provided for draining the liquid from the bottom of the precipitator.

Gas-in-Liquid Systems--
There are two general types of gas-in-liquid dispersions. The first is an unstable dispersion of gas bubbles in a liquid which is relatively simple to separate. The second type is a stable dispersion or foam that may be extremely difficult to separate. In the processing of materials that produce foam in the course of their handling, automatic systems for sensing and controlling can be installed. Instead of chemical defoamers, or in conjunction with them, physical methods may be used to accelerate the disintegration of foam. Foam structure can be attacked mechanically, thermally, or electrically. Mechanical methods use static or rotating breaker bars, rotating slingers, stationary bars or closely spaced plates. Thermal methods utilize a hot surface in contact with or near a foam to destroy the foam. Evaporation presumably removes sufficient liquid from the lamina to cause it to collapse. Electrical methods have also been used at times to weaken or destroy foam. A pair of electrodes placed in a foam mass break the foam by discharging the component bubbles.

LIQUID-SOLID SYSTEMS

Low Viscosity Mixing

A variety of process functions, including suspension or dispersion of particulate solids in a liquid to produce uniformity or to initiate and assist chemical reaction and reduction of particle agglomerate size, are carried out in vessels stirred by rotating impellers. Impellers may be divided into two broad classes, axial-flow impellers and radial-flow impellers, while tanks may be baffled or unbaffled.

Axial-Flow Impellers--
Axial-flow impellers include all impellers in which the blade makes an angle of less than 90 degrees with the plane of rotation. Propellers and pitched-blade turbines or paddles are representative axial-flow impellers. Propeller mixers are often used for agitation in tanks smaller than 1000 gallons. They may be clamped on the side of an open vessel with the shaft mounted in an angular, off-center position or bolted to a flange or plate on the top of a closed vessel. For suspension of solids, it is common to use gear-driven units, while for rapid dispersion or fast reactions, high-speed units are more appropriate. Pitched-blade turbines are used on top-entering agitator shafts instead of propellers when a high axial circulation rate is desired. A pitched-blade turbine near the upper surface of liquid in a vessel is effective for rapid submergence of floating particulate solids.

Radial-Flow Impellers--
These impellers have blades which are parallel to the axis of the drive shaft. Smaller multiblade units are known as turbines and come in a variety of types, such as curved-blade and flat-blade. Curved blades aid in starting an impeller in settled solids. Larger, slower-speed impellers, with two or four blades, are often called paddles. A paddle agitator has a diameter usually greater than 0.6 of the tank diameter and turns at a slow speed. Most large-scale agitation of solid-liquid suspensions is done with top-entering turbines or paddles.

Unbaffled Tanks--
If a low-viscosity liquid is stirred in an unbaffled tank by an axially mounted agitator, there is a tendency for a swirling flow pattern to develop. Increased vertical circulation rates may be obtained by mounting the impeller off-center. With axial-flow impellers, an angular off-center position may be used. Changes in tank size may also affect the flow pattern in such vessels. Paddle and anchor impellers normally operate coaxially within unbaffled tanks, since they may have a close clearance with the tank wall.

Baffled Tanks--
Baffles are flat vertical strips set radially along the tank wall that provide vigorous agitation of thin suspensions. Four baffles are almost adequate and a common baffle width is one-tenth to one-twelve of the tank diameter. Baffles are commonly used with turbine impellers and on-center-line axial-flow impellers and result in a large top-to-bottom circulation without vortexing. If the circulation pattern is satisfactory when the tank is unbaffled, but a vortex creates a problem, partial-length baffles may be used.

Viscous Material Mixing

Batch Mixers--
Change-can mixers are vertical batch mixers in which the container is a separate unit easily placed in or removed from the frame of the machine. The rotating elements have many forms, ranging from smooth flat blades to intricate intermeshing paddles. Stationary-tank mixers provide intense shear by maintaining very close tolerances between the mixing elements and the

housing. The gate mixer, one of the oldest stationary-tank mixers, utilizes a flat rotating structure of horizontal and vertical bars to provide shear with stationary bars fastened at the tank wall. Shear-bar mixers and helical-blade mixers are also forms of stationary-tank mixers.

The double-arm kneading mixer consists of two counterrotating blades in a rectangular trough curved at the bottom to form two longitudinal half-cylinders and a saddle section. The blades may be tangential or overlapping. Agitator blades consist of the sigma, dispersion, multiwiping overlap, single-curve, and double naben. Intensive mixers such as the Banbury can utilize heavy shafts, stubby blades, and close clearances to provide high shear mixing while those like the Prodex-Henschel and the Welex-Papenmeir combine vortex flow and high shear to achieve desired mixing. Roll mills can provide exceedingly high localized shear, while retaining extended surface for temperature control. Two-roll and three-role mills are available. Miscellaneous batch mixers include bulk blenders, the Littleford-Lodige mixer, cone and screw mixers, and pan muller mixers.

Continuous Mixers--
Single-screw extruders are frequently used as mixing devices in the plastics industry and can be equipped with large gears and thrust bearings to operate with high torque and high power input to the material. The Rietz Extractor has orifice plates and baffles along the vessel while the rotor carries multiple blades with a forward pitch, producing the head for extrusion as well as battering the material to break up agglomerates between the baffles. The Baker Perkins Ko-Kneader uses a single-screw mixer and three rows of teeth protruding inward from the barrel wall while the Transfer-mix employs a screw and barrel divided into frastoconical sections with helical channels. Twin-screw continuous mixers may be either tangential or intermeshing. The ZSK twin-screw machines are equipped with co-rotating screws which are individually made up of different screw and kneading elements slipped onto shafts. The multipurpose mixer uses pairs of agitator elements to cause alternate compression and expansion resulting in intense shear for mixing while the Farrel continuous mixer consists of rotors which act as screw conveyors to propel the feed to the mixing section where shear is provided between the rotor and chamber wall, kneading between the rotors, and the rolling action of the material itself. Miscellaneous continuous mixers consist of trough and screw mixers, pug mills, the Kneadermaster, and static mixers.

Crystallization Equipment

(See Section on Miscellaneous Separation Processes - Crystallization)

Ion-Exchange and Adsorption Equipment

(See Section on Adsorption and Ion Exchange Equipment)

Leaching

(See Section on Miscellaneous Separation Processes - Leaching)

Gravity Sedimentation Systems

Thickeners--
The oldest and simplest devices used for thickening of solids are batch settling tanks, however, where space is limited or the thickener must be housed, a filter-type mechanical thickener may be used. Owing to the difficulty in removing solids from larger batch sedimentation tanks, batch settling tanks are size limited. Consequently, continuous thickeners are generally applied in usual operations.

The basic continuous gravity thickener consists of a tank, a means for introducing the feed, a drive-actuated rake mechanism for moving settled solids to a discharge point, a means for removing the thickened solids, and a means for removing the clarified liquor. Single-compartment (unit) thickeners are available as bridge-supported thickeners, center-column-supported thickeners, and traction thickeners. Tray thickeners consist of a tank divided vertically into compartments with the feed evenly split to each compartment. Variations of this basic design are found in the washing-type, the combination, and the peripheral underflow removal tray thickeners.

Clarifiers--
Continuous clarifiers are generally employed with dilute suspensions and their primary purpose is to produce a relatively clear overflow. In clarification applications the thickened sludge produced is smaller in volume and the solids are usually lighter than that encountered with thickeners, consequently the torque requirements are lower for a clarifier than for a thickener of similar diameter. A typical rectangular clarifier utilizes a chain-type drag. The drag moves the deposited sludge to a hopper located at one end by means of scrapers fixed to the endless chains. Circular clarifiers are available in diameters of 8 to 400 ft. Similar to bridge-supported mechanism, center-column-supported mechanism with central drive, and center-column-supported mechanism with peripheral traction drive. Circular clarifiers are often equipped with a surface-skimming device, which includes a rotating skimmer, a scum baffle, and scum-box assembly. Reactor-clarifiers accomplish mixing, flocculation, and sedimentation all in a single tank. The high-rate solids-contact type is the most efficient.

Filtration Equipment

Cake Filters--
Filters that accumulate appreciable visible quantities of filtered solids on the surface of the filter medium are called cake filters. Inasmuch as the cake is itself the effective filter medium, the base on which it is deposited need not be a particularly retentive medium. For this reason, cake filtration seldom yields a completely clear filtrate.

Gravity (Hydrostatic Head) Filters--
In a gravity filter, the flow of filtrate results from the hydrostatic pressure of the column of prefilt that stands above the surface of the filter medium or the cake. Except on very small plant scale, gravity filters are seldom used in the process industries. The gravity nutsche is a tank

equipped with a false bottom, perforated or porous, that may support a filter medium or may itself act as the septum. In a gravity nutsche, the slurry contained in the tank is filtered under its own hydrostatic head with the filtrate collecting in a sump beneath the filter. Nutsches are seldom larger than 8 ft in diameter and frequently are plant-constructed from metal or wood with the false bottoms made of perforated plate, porous sintered metal, or porous ceramic slabs or blocks. The Delpark industrial filter is a semicontinuous self-cleaning gravity filter for use with freely filtering suspensions of relatively large solids. It consists of a flat endless conveyor of open screen that operates over carrying and driving rolls between sloping sides and up the ramped ends of a rack that sits above a receiving tank. The conveyor carries on its top surface a lightweight non-woven filter medium which retains the solids. Bag filters consisting of bags or pouches of filter fabric, felt, or chamois hung from suspending frames are still occasionally used. The commonest type of gravity filter is the sand or anthracite-bed filter.

Pressure Filters--

Pressure filters are those which operate under superatmospheric pressure at the filtering surface and atmospheric or greater pressure at the downstream side of the septum. Pressure filters are fed by plunger, diaphragm, screw and centrifugal pumps; blowcases; and streams that come from a pressure reactor. Continuous pressure filters now exist as well as batch or intermittent modes.

Among batch pressure filters are the pressure nutsches and sand filters, such as the upflow sand filter, in which the prefilt is pumped upward through the graded bed of sand and gravel. The filter press, the simplest of all pressure filters, is found in two basic designs: the flash-plate or plate-and-frame press, and the recessed-plate press. A plate-and-frame press is an assembly of alternate solid plates, the faces of which are studded, grooved, or perforated to permit drainage, and hollow frames, in which the cake collects during filtration. A recessed-plate filter press consists only of plates, both faces of each plate hollowed out to form a chamber for cake accumulation between each two plates. The pressure leaf filter consists of an assembly of flat filtering elements (leaves) supported vertically in a pressure shell. A filter leaf consists of a heavy screen or grooved plate over which a filter medium of woven fabric or fine wire cloth is fitted. The horizontal plate filter consists of a number of horizontal circular drainage plates and guides placed one above another in a coaxial cylindrical shell and connected in parallel. The centrifugal-discharge filter uses horizontal top-surface filter plates mounted on a hollow motorconnected shaft that serves both as a filtrate-discharge manifold and as a drive shaft to permit centrifugal removal of the cake. The industrial tubular filter consists of one or more perforated tubes supported horizontally or vertically by a transverse tube sheet within a cylindrical shell the axis of which is parallel to those of the tubes. The Burt filter is a rotating batch pressure filter lined with drainage panels covered by filter cloth. Air under pressure is admitted to the housing to force the feed through the filter.

Multicompartment drums controlled by a rotary valve and discharged by a knife are the essential elements of continuous rotary pressure filters. They are enclosed within a pressure shell and operate at substantial superatmospheric levels. The cake is removed from the pressure filter at full filtering pressure and must be throttled to atmospheric during its ejection from the filter enclosure. Continuous pressure drums are constructed in the size range 4 to 700 sq ft of filter area. The BHS-Fest filter is a continuous pressure filter that permits staged processing of the filter cake, e.g., washing and drying, and continuous discharge of the cake. Continuous precoat pressure filters are also available that operate on a long transient cycle.

Vacuum Filters--
Vacuum filters are those that operate with less than atmospheric pressure on the downstream side of the filter septum. The prefilt slurry may be fed to the filter tank by a low-head pump or by gravity. Batch and continuous filters are employed in process applications with the latter predominating. Vacuum nutsches and vacuum leaf filters are used as batch vacuum filters. The filtrate-collection sump of a nutsche may be connected to a vacuum system to convert the nutsche into a vacuum filter. The Galigher tilting filter uses a horizontal vacuum table, a shallow pan with a drainage grid and a medium support for a floor while the filter is tilted by a handwheel. The Moore filter, a vacuum leaf filter, uses a battery of leaves connected to a vacuum manifold in a portable rack. Each leaf consists of a rectangular frame of perforated pipe over which a bag of filter medium is stretched. The Sparkler HCV filter is another type of tank-mounted vacuum leaf filter. Continuous vacuum filters fall into three classes: drums, disks, and horizontal filters. They all have the following features: a filtering surface that moves from a point of slurry application, where the cake is deposited under the impetus of a vacuum, to a point of solids removal, where the cake is discharged by mechanical and pneumatic methods, and thence back to the point of slurry application.

Multicompartment Drum Filters--
The oldest and most popular multicompartment drum filter is the conventional rotary drum. The Oliver filter consists essentially of a cylindrical drum supported in an open-top tank or vat in such a manner as to allow rotation of the drum therein around its own axis in a horizontal plane. The drum shell is composed of a number of shallow compartments covered with a drainage grid and a filter cloth. The filter cake is usually discharged from the drum surface by a scraper blade. With the string-discharge filter, a system of endless strings pass around the filter drum and tangentially lift off the cake and discharge it. Removable-medium filters provide for the filter medium to be removed and reapplied as the drum rotates. Top-feed filters dewater coarse, rapidly settling solids, by feeding the slurry near the top of the drum to its ascending face, whereupon the cake is discharged at the bottom, and directed into a chute. An internal feed filter, such as the Dorrco filter, is a vacuum filter of the rotary-drum type with the filter medium placed on the inner surface of the drum as a series of panels parallel to the drum axis. The drum also serves as the container for the pulp. A major modification of the conventional continuous drum-type filter is the continuous vacuum precoat filter. The drum, with a heavy layer of filter aid formed on its surface, rotates as a thin film of solids is continuously formed on the filter aid. The rotation continues through the washing and drying zone to the discharge point where an advancing knife-edge shaves off the film of solids.

Single-Compartment Drum Filters--
These filters include the Bird-Young filter which has no internal drum piping or automatic rotary valve; the entire inside of the drum is subject to vacuum and the cake is discharged by a pulsating air blowback, without the aid of scrapers or strings. The perforated cylinder is divided into sections 2 to 2.5 in wide with the filter medium positioned into tubes between the sections and locked into place by a round rod. The filter is designed to operate at high rotative speeds.

Continuous Vacuum Disks--
The continuous rotary vacuum disk filter is offered by most makers of vacuum drum filters. It consists essentially of a number of filter disks mounted at regular intervals around a hollow center shaft. Rotation is by a gear drive. Each disk consists of trapezoidal sectors that support a filter cloth. Each sector has an outlet nipple to the center shaft and the shaft terminates in a port at an automatic valve. The assembly of filter disks on the center shaft is mounted in the feed tank so that sectors are completely submerged during the cake-building portion of the cycle. Crapers or tapered discharge rolls for each disk are mounted at the top of the tank.

Horizontal Continuous Vacuum Filters--
This class of continuous vacuum filters is characterized by a horizontal filtering surface in the form of a table, a belt, or multiple pans in linear or circular arrangement. The horizontal table filter uses a rotating annular table whose top surface is a filter medium. The table is divided into sectors and a vacuum is applied through a drainage chamber beneath the table that leads directly into a large rotary valve. Slurry is pumped onto the table and the cake is removed by a scroll conveyer which elevates it over the side of the filter. The tilting-pan filter makes each sector into a pan connected by a radial arm to a central vacuum valve. The pan is carried on a roller that rides a circular track around the filter and at the point of cake discharge a mechanism inverts the pan. Another type is the horizontal belt filter that uses a slotted or perforated endless elastomer belt supporting a filter fabric and traveling across a suction box. The slurry is pumped onto the filter at one end and the cake is dumped at the other end.

Filter Thickeners--
Thickeners are devices which remove a portion of the liquid from a slurry to increase the concentration of solids in suspension. A common method of thickening is to use gravity sedimentation tanks.

The Peterson Roto-Disc clarifier is an intermittent totally submerged continuous vacuum disk filter with conventional segmented disks and valving for either vacuum or back pressure. The disk filters until it has accumulated cake of suitable thickness, whereupon the cake is discharged into the tank and the cycle is repeated. The cake settles rapidly to the conical bottom of the tank, where it can be pumped away continuously. Another filter-type thickener, the Shiver continuous thickener, is a modified filter press with special plates instead of frames. The special plates carry spiral or vertical channels to feed liquid from left to right across the plate by successive traverses. As the slurry flows across the plate, part of the suspending liquid is filtered through a cloth to an adjacent drainage plate allowing the thickened slurry to emerge from the press continuously.

Clarifying Filters--
Clarifying filters are also available, such as disk filters and plate presses in which the liquid flows through the disks and into a central or peripheral discharge manifold. Ultrafilters use a media of sheets of fiber and the filtration is normally performed in single-sheet cells or multiplate presses. Precoat pressure filters consist of one or more leaves, plates, or tubes upon which a coat of diatomaceous earth or other filter aid is deposited to form a filtering surface for clarification. Cartridge clarifiers are units which consist of one or more replaceable or renewable cartridges containing the active filter element.

Centrifuges

Sedimentation Centrifuges--
Sedimentation centrifuges remove or concentrate particles of solids in a liquid by using the particles to migrate through the fluid radially toward or away from the axis of rotation. The tubular-bowl centrifuge consists of a bowl suspended from an upper bearing and drive assembly through a flexible-drive spindle. Feed enters the bottom of the bowl through a stationary feed nozzle under pressure. The incoming liquid is accelerated, moves upward through the bowl as an annulus, and discharges at the top. Solids that have sedimented against the bowl wall are removed manually. A multichamber centrifuge has a bowl that consists of a series of short tubular sections of increasing diameter nested to form a continuous tubular passage for the flow of liquid. The heaviest particles are deposited in the smallest-diameter tube and smaller, lighter particles in the larger-diameter zones. Disk centrifuges admit feed to the center of the bowl near its floor and it rises through a stack of sheet-metal "disks", each with several holes which form channels when the disks are stacked. Solid particles are removed from the liquid and deposited on the wall of the bowl. Peripheral-discharge disk centrifuges use bowl walls that are sloped to direct the sedimented solids to a narrow annulus at the periphery. Continuous decanter centrifuges consist of a solid-wall bowl with a horizontal or vertical axis of rotation. The solids-discharge ports at one end of the bowl are conventionally at a smaller radius than the liquid-discharge ports at the other end. A helical screw conveyor continuously transports the heavy solids that have sedimented against the bowl wall. Knife-discharge centrifugal clarifiers with solid instead of perforated bowls are used as sedimentation clarifiers. The feed enters the bowl at the hub end and the clarified effluent either overflows the lip ring or discharges through a skimmer pipe. The solids are out, as from centrifugal filters.

Centrifugal Filters--
Centrifuges that filter, i.e., cause liquid to flow through a bed of solids held on a screen, are commonly called centrifugal filters or centrifugals. Variable-speed basket centrifuges rotate on a vertical axis and the basket is usually perforated and cylindrical in shape. It is connected to the drive shaft through a hub. The hub may be solid or open. Unloading is through the

bottom, either manually while the basket is at rest, or by an unloader knife while the basket is slowly rotating. Constant-speed basket centrifuges operate at a constant bowl speed during the entire sequence. Constant-speed batch-automatic centrifuges practically always operate on a horizontal axis of rotation. After the cake has been spun dry, the cake is peeled out by rotating an unloader knife into it. Continuous filtering centrifuges are available as are conical-screen centrifugals and pusher (reciprocating) centrifugal filters.

Expression Equipment

Expression is the separation of liquid from a two-phase solid-liquid system by compression under conditions that permit the liquid to escape while the solid is retained between the compressing surfaces. Expression equipment can be batch or continuous.

Batch Presses--
The principal batch presses are the box, platen, pot, curb, and cage. In the box press, the material to be expressed is wrapped in canvas cloths and placed in a series of steel boxes fitting between the fixed and movable heads of a vertical hydraulic press. The platen press is similar to the box press, but the cloth bags are not enclosed on the sides during pressing. Material to be pressed in a pot press is enclosed in a cylindrical pot, with filter pads or screens beneath and above, and is compressed by a ram entering from above. In the curb press, material to be expressed is enclosed in a cylinder of wooden slats or beveled steel bars or perforated steel plates. Compression by a ram causes the liquid to escape through the walls of the cylinder and flow to collecting channels at the base. The cage press is similar to the curb press except that the inside of the cylinder has fine longitudinal grooves leading through the cylinder walls to larger drainage channels.

Continuous Presses--
The screw press and various types of roller mills are in common use. The continuous screw press consists of a rotating screw fitting closely inside a slotted or perforated curb. The discharge end of the curb is partly closed by an adjustable cone to change the size of the opening and thus vary the pressure on the material. Rotation of the screw moves the material forward and, as the pressure increases, liquid is expelled. Continuous roller mills combine a mechanical breaking and crushing action with pressure to express juice. Three-roll mills are common, with the top roll above and between the other two. Material is squeezed between the top and first rolls and then is directed by a turnplate into the nip of the top and second rolls for a second pressing. Continuous disk presses are also in use for the mechanical removal of moisture from various materials.

GAS - SOLID SYSTEMS

Solids - Drying Equipment

A classification of drying systems based on the method of transferring heat to the wet solids is found in Figure A-1. Various types of dryers are listed and brief descriptions are given.

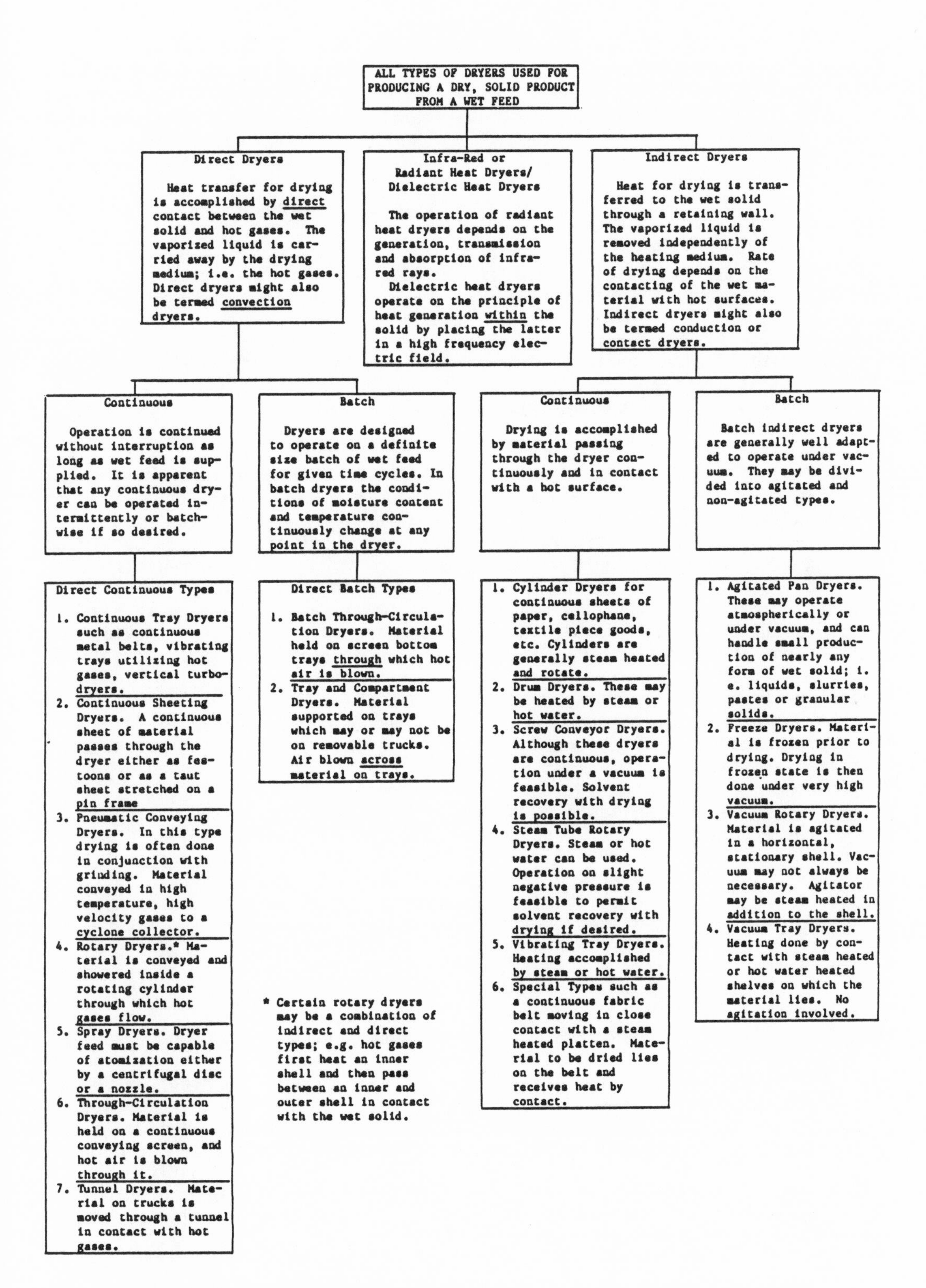

Figure A-1. Classification of dryers, based on method of heat transfer.

Fluidized-Bed Systems

The possible uses of fluidized beds are manifold. Generally, the fluidized bed is used for gas-solids contacting; however, in some cases the presence of the gas or solid is used only to provide a fluidized bed to accomplish the end result. The greatest commercial uses of fluidized beds are in petroleum cracking (catalytic chemical reactors) and in roasting certain ores (noncatalytic heterogeneous reactors).

The major parts of a fluidized-bed system that would be in contact with hazardous fluids can be listed as follows:

1. Reaction vessel
2. Solids feeder or flow control
3. Solids discharge
4. Dust separator for the exit gases

Reaction Vessel--

The process of fluidizing converts a bed of solid particles into an expanded suspended mass that resembles a boiling liquid. The usual shape of the reaction vessel is a vertical cylinder. Refactory lined steel is usually used. A reinforced gunnite lining has been found satisfactory for catalytic cracking of petroleum. Ample foundations and a sturdy supporting structure are required for the reactor. The reactor must also include freeboard or disengaging height between the top of the fluid bed and the gas-exit nozzle. A gas distributor (a number of variations are in use) is also provided to achieve good distribution of the gas and to prevent backflow of solids.

Solids Feeder--

In the case of catalytic-cracking units the makeup catalyst may be fed from pressurized hoppers into one of the conveying lines. The pressure is increased at the bottom of the solids draw-off pipe or standpipe by introducing gas at the bottom or at intervals along the length. Several designs of valves for solids flow control are used: slide valve, star valve, table feeder, screw feeder, and cone valve. Usually, block valves are used in conjunction with the control valves. Where the solid is one of the reactants, the flow must be continuous and precise to maintain constant conditions in the reactor. Standard commercially available solids conveying equipment can be used to control the rate and deliver the solids to the feeder. Screw conveyors, dip pipes, seal legs, and injectors are used to introduce the solids into the reactor proper.

Solids Discharge--

The type of discharge mechanism utilized is dependent upon the necessity of sealing the atmosphere inside the fluidized-bed reactor and the subsequent treatment of the solids. The simplest solids discharge is an overflow weir. A simple flapper valve is frequently used to restrict the flow of gas through the opening. Overflow to combination seal and quench tanks is used where it is permissible to wet the solids and where disposal or subsequent treatment of the solids in slurry form is desirable. Seal legs are frequently used in conjunction with solids-flow control valves to equalize pressures and to strip trapped or adsorbed gases from the solids. The FluoSeal is a simple and effective way of sealing and purging gas from the solids where an overflow-type discharge is used.

Dust Separation--
It is usually necessary to recover the solids carried by the gas leaving the disengaging space or freeboard of the fluidized bed. Generally, cyclones are used to remove most of these solids. Fluidized-bed cyclone arrangements are: (a) single-stage internal, (b) two-stage internal, (c) single-stage external, dust returned to bed, (d) two-stage external, dust returned to bed, and (e) two-stage external, dust collected externally. Since cyclones are less effective as the particle size decreases, secondary collection units are frequently required, i.e., filters, electrostatic precipitators, scrubbers. On small-scale units filters can be used without the use of cyclones to reduce the loading of solids in the gas. Multiple units must be provided so that one unit can be blown back with clean gas while one or more are filtering.

Gas - Solids Separation Equipment

Gravity Settling Chambers--
The gravity settling chamber, the simplest and earliest type of dust-collector equipment, consists of a chamber in which the gas velocity is reduced to enable dust to settle out by the action of gravity. Its industrial utility is limited to removing particles larger than 43 microns diameter. Gravity collectors are generally built in the form of long, empty, horizontal, rectangular chambers with an inlet at one end and an outlet at the side or top of the other end. The Howard dust chamber uses horizontal plates arranged as shelves within the chamber to give a marked improvement in collection.

Impingement Separators--
When a dust-laden fluid impinges on a body, the fluid will be deflected around the body, whereas the dust particles will tend to be collected on the surface of the body. A typical commercial impingement collector is the reverse-nozzle impingement separator. In general, impingement collectors are designed for a pressure drop in the range of 0.1 to 0.5 in water and are limited to removing dusts that are predominantly larger than 10 to 20 microns diameter.

Cyclone Separators--
The most widely used type of dust-collection equipment is the cyclone, in which dust-laden gas enters a cylindrical or conical chamber tangentially at one or more points and leaves through a central opening. The immediate entrance to a cyclone is usually rectangular. The dust particles, by virtue of their inertia, will tend to move toward the outside separator wall from which they are led into a receiver. Typical commercial cyclones are: (a) the Duclone collector, (b) Sirocco type D collector, (c) Van Tongeren cyclone, (d) Multiclone collector, (e) the Dustex miniature collector assembly, and (f) the rotational-flow centrifugal separator.

Mechanical Centrifugal Separators--
A number of collectors are commercially available in which the centrifugal field is supplied by a rotating member. Typical mechanical centrifugal separators are the Type D Rotoclone and the Sirocco cinder fan. In the former, the exhauster or fan and dust collector are combined as a single unit. The blades are especially shaped to direct the separated dust into an annular slot leading to the collection hopper while the cleaned gas continues to the scroll outlet. In the latter, the unit is usually used on the inlet side of a fan with the rotor connected to the fan shaft. The dust-laden gas enters on the periphery of the scroll, passing radially inward through the rotor and out the center. Dust thrown to the scroll wall is concentrated in a small stream of gas which is by-passed through a cyclone collector, where the dust is finally gathered.

Granular-bed Separators--
One of the earliest of these devices is the Lynch granular filter. It utilizes a packed bed of gravel which is withdrawn from the bottom of the bed continuously and passed over a screen to remove collected dust before the gravel is returned to the top of the bed. The Ducon expandable-bed filter with cylindrically or conically shaped baffles containing the bed granules is available commercially. The filter elements are arranged within a cylindrical or rectangular housing similarly to cloth tubes in bag houses. The granules are retained between two concentric perforated tubes. The inlet gas passes through the outer perforated tube and down through the granular fixed beds which collect the solids. A periodic reverse flow of high-pressure air removes the solids from the fixed bed to a dust hopper.

Bag Filters--
There are two general types of bag filters. The older type employs a relatively thin woven fabric as the filter medium while the other uses felt. In the operation of woven-fabric filters, the dust-laden gases are passed through a woven fabric which "filters" out the dust, allowing the gases to pass on. When the dust-laden gases first pass through the cloth, the efficiency of separation will be low until enough particles have been removed to build up what corresponds to a "precoat" in the fabric pores. Mechanical filters, available as standard commercial units, comprise two general types. One utilizes cloth envelopes supported by screens and the other uses either oval or round vertically mounted bags. Felt-fabric filters permit somewhat higher dust-laden air velocities without excessive penetration, since the filter cake plays less of a role in filtration with this type of collector than it does with the woven-fabric filters. Typical felt-fabric filters include: (a) the Hersey reverse-jet filter, (b) unhoused reversed-jet filter employing felt bags, (c) reverse-jet filter cleaned by induced flow of exit gas, and (d) the Dustex Inductaire bag filter, which may be used with felt or woven-fabric bags.

Scrubbers--
Scrubbers remove dusts and mists from gases, in which a liquid, usually water, is added or circulated to assist in the collection process. A wide variety of units are commercially available. The simplest type of scrubber is a spray tower in which liquid droplets are produced by spray nozzles and are allowed to settle through the rising gas stream. Conventional full-cone spray nozzles are commonly used. Cyclone spray scrubbers take advantage of

the centrifugal force in a spinning gas stream. Impingement scrubbers use impingement plates consisting of perforated sheets and impingement baffles located above each perforation. Packed- and fluidized-bed scrubbers utilize turbulent conditions, as the gas enters below the bed and passes upward, to clean the gas and minimize fouling. Orifice scrubbers achieve high aerosol velocities by forcing the gas through restrictions of various shapes. Venturi scrubbers are commercially available and have wide application in the collection of micron and submicron aerosols. Water-jet scrubbers use the action of the scrub water to provide motive power for the gas. In mechanical scrubbers, water is sprayed into the inlet of a fan where the fan blades atomize the liquid and accelerate it to high velocity so that inertial impaction takes place between the dust particles and the liquid drops. Also, fibrous-bed scrubbers have been developed for fine dusts, mists, and fumes.

Electrical Precipitators--
When particles suspended in a gas are exposed to gas ions in an electrostatic field, they will become charged and migrate under the action of the field. There are two general classes of electrical precipitators: single-stage, in which ionization and collection are combined and two-stage, in which ionization is achieved in one portion of the equipment, followed by collection in another. The single-stage type of unit, commonly known as a Cottrell precipitator, is most generally used for dust or mist collection from industrial process gases. The corona discharge is maintained throughout the precipitator, providing initial ionization and preventing redispersion of precipitated dust. Cottrell precipitators may be divided into two main classes: the so-called plate type in which the collecting electrodes consist of parallel plates, screens, or rows of rods, chains, or wires; and the pipe type in which the collecting electrodes consist of a nest of parallel pipes which may be square, round or any other shape. In two-stage precipitators, corona discharge takes place in the first stage between two electrodes having a non-uniform field. This is generally obtained by a fine-wire discharge electrode and a large-diameter receiving electrode. The second-stage involves a relatively uniform electrostatic field in which charged particles are caused to migrate to a collecting surface. This stage usually consists of either alternately charged parallel plates or concentric cylinders with relatively close clearances compared with their diameters.

Air Filters--
Air filters are employed in the elimination of atmospheric dust. No attempt to recover the dust is usually made. Air filters may be classified in three groups on the basis of the type of filter medium employed: viscous, dry, or automatic. Viscous filters are so called because the filter medium is coated with a viscous material to retain the dust. The filter pad may consist of glass fibers, animal hairs, wood shavings, corrugated fiberboard, split wire, or metal screening. These are coated with a dust-collecting liquid, such as mineral oil. Both reusable and disposable viscous-type filters are available. Dry filters are supplied in units similar in size to the viscous filters, except that the depth is usually greater. The filter mediums are generally sheets of cellulose pulp, cotton, flet or spun glass. Filters using felt or similar materials are generally reconditioned by vacuum or dry cleaning.

Automatic filters are so called because the cleaning operation is essentially continuous and automatic. Most commercial automatic filters are of the viscous type and consist of perforated, crimped, or woven metallic screens in series. The screen curtains are drawn around in a vertical direction, either continuously or intermittently. An oil bath serves to rinse out the dust and coat the screen with a fresh film of oil. The dust is then allowed to settle out as a sludge in the bottom of the hopper. The Airmat dust arrestor is a dry automatic filter when it is applied to dusts that are relatively non-sticky and easily shaken off. The air flow must be diverted, however, when the filter is vibrated.

LIQUID-LIQUID SYSTEMS

Stagewise Equipment

The function of a stage is to contact the liquids, allow equilibrium to be approached, and to make a mechanical separation of the liquids. The contacting and separating correspond to mixing the liquids and settling the resulting dispersion such that these devices are usually called mixer-settlers. The operations may be carried out in batch fashion or with continuous flow. If batch, it is likely that the same vessel will serve for both mixing and settling, whereas if continuous, separate vessels are usually but not always used.

The equipment for extraction or chemical reaction may be classified as follows:

A. Mixers

1. Flow or line mixers

a) Mechanical agitation

b) No mechanical agitation

2. Agitated vessels

a) Mechanical agitation

b) Gas agitation

B. Settlers

1. Non-mechanical

a) Gravity

b) Centrifugal (cyclones)

2. Mechanical (centrifuges)

3. Settler auxiliaries

    a) Coalescers

    b) Separator membranes

    c) Electrostatic equipment

In principle, at least, any mixer may be coupled with any settler to provide the complete stage. There are several combinations which are especially popular. Continuously operated devices usually, but not always, place the mixing and settling functions in separate vessels. Batch-operated devices may use the same vessel alternately for the separate functions.

Flow or Line Mixers--
Flow or line mixers are devices through which the liquids to be contacted are passed and are used only for continuous operations or semibatch (where one liquid flows continuously and the other is continuously recycled). If holding time is required for extraction or reaction, a vessel of the necessary volume may be a long pipe of large diameter, sometimes fitted with segmental baffles, but frequently the settler which follows the mixer serves.

There are many types, and only the most important will be mentioned here. Jet mixers depend on impingement of one liquid on the other to obtain a dispersion. One of the liquids is pumped through a small nozzle or orifice into a flowing stream of the other. Injectors are flow mixers in which the flow of one liquid is induced by the flow of the other, with only the majority liquid being pumped at relatively high velocity. Orifices and mixing nozzles can provide constrictions in a pipe through which both liquids are pumped resulting in a pressure drop that is partly utilized to create a dispersion. Single nozzles or several in series may be used. Valves may be considered as adjustable orifice mixers, while centrifugal pumps, where two liquids are fed to the suction side of the pump, have been used fairly extensively for mixing. An agitated line mixer, which combines the features of orifice mixer and agitator, is used extensively in treating petroleum and vegetable oils.

Also, co-current flow of immiscible liquids through a packed tube produces a one-stage contact, characteristic of line mixers, while a pipeline can be used to disperse one liquid in another as they flow co-currently through the pipe.

Agitated Vessels--
Agitated vessels may frequently be used for either batch or continuous service, and for the latter may be sized to provide any holding time required. Mechanical agitation utilizes a rotating impeller immersed in the liquid to accomplish the mixing and dispersion. Hundreds of devices are available using this principle. The basic requirements regarding shape and arrangement of the vessel, type, and arrangement of the impeller, and the like, are essentially the same as those for dispersing finely divided solids in liquids. A baffled mixing vessel is a typical single-compartment vessel

for extraction or chemical reaction. Back-mixing may be reduced, and the extraction stage efficiency thereby increased, by dividing the vessel into a number of compartments by horizontal separators, each compartment containing an agitator.

Settlers--
The mixture of liquids issuing from a mixer must be settled, coalesced, and separated into its constituent liquid phases in bulk in order to withdraw the separated liquids from a stage. Gravity settlers or decanters are tanks wherein the dispersion is continuously settled and coalesced, and from which the settled liquids are continuously withdrawn. Typical settlers include types, such as: (a) simple horizontal, (b) vertical, (c) baffled, and (d) cone-bottom. Cyclones have been tried as a simple means of enhancing, by centrifugal force, the rate of settling of liquid dispersions, but the results were not promising. Mechanical centrifuges, high-speed machines, have been used for many years for separating liquid-liquid dispersions. Settler auxiliaries include coalescers, separating membranes, and electrical devices.

Continuous Contact Equipment

Equipment in this category is usually arranged for multistage counter-current contact of the insoluble liquids, without repeated complete separation of the liquids from each other between stages.

Equipment can be broadly classified into the following categories, generally in order of increasing complexity of internal construction. Those most generally used are:

A. Gravity-operated extractors

1. No mechanical agitation

a) Spray towers

b) Packed towers

c) Perforated-plate (sieve-plate) towers

2. Mechanically agitated extractors

a) Towers with rotating stirrers

b) Pulsed towers

o Liquid contents pulsed

o Reciprocating plates

B. Centrifugal extractors

Spray Towers--
These are simple gravity extractors, consisting only of empty towers with provisions for introducing and removing the liquids at the ends. Because of

severe axial dispersion, it is practically very difficult to obtain the equivalent of more than one or two theoretical stages or transfer units. Consequently, spray towers have largely been abandoned for industrial liquid extraction, although in recent years considerable interest has developed in their use for direct-contact heat exchange, where more than two transfer units may not be necessary. Ordinarily, straight-sided towers are used. The orifices or nozzles for introducing the dispersed phase are usually not smaller than 0.05 in in diameter in order to avoid clogging, nor larger than 0.25 in in order to avoid formation of excessively large drops. Also they should be designed to eliminate wetting by the dispersed liquid.

Packed Towers--
For a packed tower, the empty shell of a spray tower is filled with packing to reduce the vertical circulation of the continuous phase. The standard commercial packings used in gas absorption, such as Raschig rings and Berl saddles, are generally used. The standard commercial packing pieces should be no larger than one-eighth the tower diameter. The packing support is most conveniently an open bar grid. The packing reduces the flow capacity of the tower but also materially reduces the height required. Industrial applications generally use straight-sided towers. The dispersed phase is introduced through nozzles which are embedded at least 1 or 2 in into the packing. Packed towers are not particularly useful when suspended solids are involved, but for clear solutions they are used in all aspects of commercial liquid-extraction separations.

Perforated-Plate Towers--
In the most common design of the perforated-plate or sieve-plate tower, the light liquid flows through the perforations of each plate and is thereby dispersed into drops which rise through the continuous phase. The continuous liquid flows horizontally across each plate and passes through the downspout to the plate beneath. Extraction rates are enhanced by the repeated coalescence and redispersion into droplets of the dispersed phase. So the dispersed phase issues cleanly from the perforations, the material of the plates should be preferentially wet by the continuous phase or the dispersed phase should issue from nozzles projecting beyond the plate surface. Perforations are usually 1/8 to 1/4 in diameter, set 1/2 to 3/4 in apart, on square or triangular pitch. Downspouts are best set flush with the plate from which they lead, with no weir as in gas-liquid contact.

Mechanically Assisted Gravity Devices--
A rotary-disc contactor consists of a tower formed into compartments by horizontal doughnut-shaped or annular baffles and within each compartment agitation is provided by a rotating, centrally located, horizontal disk. The rotating disk is smooth and flat and of a diameter less than that of the opening in the stationary baffles. Towers have been used ranging up to 8 ft diameter by 40 ft tall. The Mixco Lightnin CMContactor is an extractor that extends the simple baffled mixing vessel into a multi-stage column. Commercial application has been made with towers of small diameter. Schiebel columns are found in two designs. The earlier model employs alternate compartments agitated with centrally located impellers while the others are packed with an open woven wire mesh. There are neither horizontal compartment plates nor vertical baffles. This was one of the first

mechanically agitated extractors used industrially in a wide variety of services. The newer design uses an impeller surrounded by a shroud baffle. The Treybal Extractor is an adaptation of a mixer-settler cascade to column form, wherein the liquids are permitted to settle completely between stages. The arrangement of the baffles in the mixing zones provide an impeller pumping action leading to enhanced flow capacities. Pulsed columns are extractors wherein a rapid reciprocating motion of relatively short amplitude is applied to the liquid contents. They may consist of ordinary extractors (spray, packed, etc.) on which pulsations are imposed, such as by a reciprocating plunger or piston pump, or a special sieve-plate design. In the latter case, the tower is fitted with horizontal sieve plates which occupy the entire cross section of the column. Controlled cycling and reciprocating plate columns are two other techniques used to improve extraction efficiencies.

Centrifugal Extractors--
Centrifugal machines are especially useful for handling liquids of low density difference and those with tendencies to form emulsions. The Polbielnick extractor uses a cylindrical drum containing concentric perforate cylinders. Liquids are introduced through a rotating shaft with the aid of special mechanical seals; the light liquid is led internally to the drum periphery and the heavy liquid to the axis of the drum. Rapid rotation causes radial counterflow of the liquids which are then led out through the shaft. Other centrifugal devices include the Quadronics extractor, the Luwesta extractor and the DeLaval extractor.

SOLID-SOLID SYSTEMS

Solids Mixing

There are several types of solids-mixing machines. In some machines the container moves. In others, a device rotates within a stationary container. A combination of rotating container and rotating internal device can also be employed. Sometimes baffles or blades are present in the mixer. Table A-4 classifies solids-mixing machines via various characteristics.

Solids Sampling

There are several types of standard mechanical samplers and many more non-standard units which have been developed for limited special applications. Some of the more common mechanical sampler types are: automatic, Snyder, and Vesin. The automatic straight-line samplers offer the lowest cost and greatest flexibility of operation and are available in both normal- and heavy-duty construction. The Snyder and Vesin-type samplers are arc-type samplers which normally operate continuously and are for sampling dry materials.

Screening

Screening is the separation of a mixture of various sizes of grains into two or more portions by means of a surface acting as a multiple go and no-go gage. The screening surface may consist of woven wire, silk, or plastic cloth; perforated or punched plate; grizzly bars, or wedge wire sections. Screening machines may be divided into five main classes: grizzlies, revolving screens,

Table A-4. Types of Solids-mixing Machines

Tumbler	Tumbler with internal agglomerate breaker	Stationary shell or trough	Both shell and internal device rotate	Impact mixing	Process steps which can affect solids mixing
Without baffles: Drum, either horizontal or inclined	Ball mill Pebble mill	Ribbon	Countercurrent, muller turret and pan rotate in opposite directions	Hammer mill	Filling of hoppers
Double cone	Rod mill	Stationary pan, rotating muller turret		Impact mill	Fluidization
Twin shell	Vibratory pebble mill	Vertical screw	Planetary types	Cage mill	Screw feeders
Cube	Double cone	Single rotor		Jet mill	Conveyer-belt loading
Mushroom type	Twin shell	Twin rotor		Attrition mill	Elevator loading
	Cube	Turbine			Pneumatic conveying
		Paddle mixer			Vibrating
With baffles: Horizontal drum Double cone revolving around long axis		Sifter (turbosifter)			

shaking screens, vibrating screens, and oscillating screens. Grizzly screens consist of a set of parallel bars held apart by spacers at some predetermined opening. A grizzly is widely used before a primary crusher in rock- or ore-crushing plants to remove the fines before the ore or rock enters the crusher. It can be a stationary set of bars or a vibrating screen. Revolving screens or trommel screens consist of a cylindrical frame surrounded by wire cloth or perforated plate, open at both ends, and inclined at a slight angle. The material to be screened is delivered at the upper end and the oversize is discharged at the lower end. Mechanical shaking screens consist of a rectangular frame, which holds wire cloth or perforated plate and is slightly inclined and suspended by loose rods or cables, or supported from a base frame by flexible flat springs. The frame is driven with a reciprocating motion as material is fed in at the upper end. Vibrating screens can be divided into two main classes: (1) mechanically driven, and (2) electrically driven. The most versatile mechanical vibration is generally conceded to be the vertical circle produced by an eccentric or unbalanced shaft. Screening machines actuated by rotating unbalanced weights have a symmetrical shaft through the screen body with an unbalanced flywheel on each end. Electrically vibrated screens are particularly useful in the chemical industry. Typical of these is the Hum-mer screen used prevalently in the fertilizer industry for handling mixed chemical fertilizers. Oscillating screens are characterized by low speed oscillations in a plane essentially parallel to the screen cloth.

Classification

Wet classification is the art of separating the solid particles in a mixture of solids and liquids into fractions according to particle size or density by methods other than screening. In general the two products are: (1) a partially drained fraction containing the coarse material (called the sand), and (2) a fine fraction along with the remaining portion of the liquid medium (called the overflow). The classifying operation is carried out in a pool of fluid pulp confined in a tank arranged to allow the coarse solids to settle out, whereupon they are removed by gravity, mechanical means, or induced pressure. Solids which do not settle report as overflow from the pool. Classifier types fall into three basic categories: (1) non-mechanical, (2) mechanical, and (3) hydraulic. Identification and pertinent data on wet classification machines are given in Table A-5.

Jigging

A jig is a mechanical device used for separating materials of different specific gravities by the pulsation of a stream of liquid flowing through a bed of materials. The liquid pulsates causing the heavy material to work down to the bottom of the bed and the lighter material to rise to the top. Each product is then drawn off separately. Jigging is best suited for coarse material that is unlocked in the size range 20 mesh and coarser and where there is a considerable difference between the effective specific gravity of the desired product and the waste material.

Table A-5. Specifications and Major Applications of Wet Classification Machines

Type of Classifier	Normal Mesh of Separation Range*	Normal Feed Tonnage Range	Maximum Oversize in Feed	Normal Overflow, % Solids Range	Typical Applications
NON-MECHANICAL:					
Cone classifer	28-325	2-100 tons/hr	1/4 in	5-30	For desliming and primary dewatering.
Liquid cyclone	35 mesh to 5u	1/2-1500 gal./min	14-325 mesh	5-30	For medium or fine separations and closed-circuit grinding.
MECHANICAL:					
Drag classifier	28-200	5-350 tons/hr	1 1/2 in	5-30	For desliming, conveying, and closed-circuit grinding.
Rake and spiral classifiers	20-200	5-350 tons/hr	1 in	5-30	Closed-circuit grinding, washing & dewatering, desliming, process feed control.
Bowl classifier	100-325	5-200 tons/hr	1/2 in	5-25	Closed-circuit grinding usually in secondary circuits.
Bowl desiltor	100-325	5-250 tons/hr	1/2 in	1-15	Recovery of fine sand, limestone, coal & fine phosphate rock from large flow volumes.
Hydroseparator	100-325	5-700 tons/hr	1/4 in	1-20	For fine separation where large feed volumes are involved and drainage not critical.
Solid-bowl centrifuge	200 mesh to 1u	10-600 gal/min	1/4 in	1-40	For fine-size fractionating.
Countercurrent classifier	20-100	1-600 tons/hr	3 in	5-30	Sand-slime separations, washing, closed-circuit grinding.
HYDRAULIC:					
Jet sizer	8-150	2-100 tons/hr	3/16 in	1-10	Multiproduct unit for exceptionally clean sands fractionated into narrow size ranges. Min. 3 tons hydraulic water per ton sand.
Supersorter**	8-150	40-150 tons/hr	3/8 in	1-10	Multiproduct unit for exceptionally clean sands fractionated into narrow size ranges. Min. 3 tons hydraulic water per ton sand.
Siphonsizer***	14-150	1-100 tons/hr	1 in	1-10	Two-product unit efficient for desliming and exceptionally clean sands, washing, closed-circuit grinding. Min. 2 tons hydraulic water per ton sand.

* Size of screen retaining 1.5% of the overflow solids.
** Trademark of Deister Concentrator Co., Inc.
*** Trademark of Dorr-Oliver, Inc.

There are two principal types of jigs. In the first type, the sieve is stationary and water is forced up through the screen. A form of the fixed-sieve-type jig is the Jeffrey air-operated Baum jig used extensively in coal washing. Pulsations are caused by applying and exhausting air pressure from the pulsation chamber. Such jigs customarily are built with a number of compartments, where each compartment or cell rejects waste material. In the second type of jig, such as the Hancock jig, the sieve moves up and down.

Tabling

Tabling is a concentration process whereby a separation between two or more minerals is effected by flowing a pulp across a riffled plane surface inclined slightly from the horizontal, differentially shaken in the direction of the long axis and washed with water flowing at right angles to the direction of motion. A separation between two or more minerals depends mainly on the difference in specific gravity between the minerals and to a lesser extent on the shape and size of the particles. The heaviest particles in a table feed are the least affected by the current of water washing down over the table, and they collect in the riffles along which they move to the end of the table. Tables are usually surfaced with either heavy linoleum or with rubber. Common practice is to use multiple decks.

Tabling may be done dry as well as wet and for such use tables of special design are used. The Sutton, Steele, and Steele table, an example of this type of equipment, uses a shaking motion somewhat similar to that of a wet table, except that the direction of motion is inclined upward from the horizontal. A blast of air serving as the distribution medium is driven through a perforated deck. This table has the ability to handle material coarser than that treated on most wet tables.

Spiral Concentration

Spiral concentration of ores and industrial materials is based primarily on the specific gravity differentials of the materials to be separated and the shape factor of the feed material. The best known of these concentrators is the Humphreys spiral concentrator which is a spirally shaped channel or launder with a modified semicircular cross section. As the feed slurry flows down the spiral channel, the particles with the highest specific gravity sink to the bottom and move inward toward the inside of the channel. The lighter-weight particles move to the outside and are carried away by the faster, more dilute pulp stream. The spiral provides repeated washing stages as the pulp flows down the channel. Generally the richest concentrate is withdrawn from the concentrate ports near the top end of the spiral. The spiral concentrator is particularly capable in the gravity processing of tailing streams from conventional magnetic and froth flotation types of ore-processing plants.

Dense-Medium Separation

Dense-medium separation is applicable to any ore in which the constituents have a measurable gravity difference. Finely divided high-gravity solids in liquid suspension produce a medium in which separation of solids can be effected. In coarse-dense medium work, three basic separatory-vessel types are most commonly used: (1) cone-type, (2) drum-type, and (3) trough-type.

The cone- and drum-type vessels have been used extensively in treating high gravity solids such as iron ore. The trough-type vessel is used in coal treatment. Specific devices include the dense-media cyclones and the Dyna Whirlpool Separator.

Magnetic Separation

Any solid placed in a magnetic field is affected by it in some way. Solids may be classified into two broad categories: (1) diamagnetic solids that are repelled, and (2) paramagnetic solids that are attracted by a magnetic field. The art of separating one solid from another by means of a magnetic field is called magnetic separation. The principal uses of magnetic separators are: (1) tramp-iron removal, and (2) concentration and purification. Tramp-iron magnetic separators are used to protect handling and processing equipment such as crushers and pulverizers. They are usually applied on dry material or on material which contains only surface moisture. The most common of these separators are magnetic-head pulleys, used to remove tramp iron from products handled on belt conveyors, suspended magnets, magnetic drums, plate magnets, grate magnets and metal detectors.

Special magnetic separators have been designed for concentration and purification of solids. Generally, concentration involves magnetic separation of a large amount of magnetic feed product, whereas purification consists of removal of small amounts of magnetic particles from a large amount of non-magnetic feed material. Magnetic separators and purifiers can be divided into: (1) wet types and (2) dry types. Three of the most frequently applied wet magnetic separators are: (1) permanent and electromagnetic-drum separators, (2) permanent and electromagnetic filters, and (3) high-intensity wet magnetic separators. Dry magnetic separators can be classified by magnetic field intensity. They include: (1) high intensity, (2) moderate intensity, (3) low intensity, and (4) high-speed low-intensity dry-drum separators.

Electrostatic Separation

Electrostatic separation is based on imparting a surface charge on one or more materials in a granular mixture before the mixture enters an electrostatic field, wherein the grains of that material will be repelled or attracted by the electrodes. By causing such grains to fall into separate chutes, a separation or concentration results. The major electrification mechanisms are: (1) contact, (2) conductive induction, and (3) ion bombardment. The two main types of equipment in use are the plate type and the rotor type. The plate type consists of two vertical plates suspended relatively close to each other and having either a high voltage impressed on one and the other grounded, or one charged positively and the other negatively, with chutes below to collect the various products. Usually the surfaces of the substances to be separated must be pretreated to facilitate and enhance particle-to-particle charging. The rotor type, conversely, is a machine that actually performs the charging step during its operation and therefore is not completely dependent on prior treatment. The rotor is used to expose the particles to an ionic charge and then remove them while simultaneously acting as a continuous grounded surface.

Flotation

Froth flotation is the dominant process of mineral dressing in use today. Mineral dressing is the treatment of ores at or near the mine site to produce concentrates of valuable minerals and tailings of wastes. The most widely used machines for sulfide, coal, and non-metallic flotation, are the Fagergren, the D-R Denver, and the Agitair flotation machines. These machines provide mechanical agitation and aeration by means of a rotating impeller on an upright shaft. Also, the Agitair and Denver units utilize air from a blower to help aerate the pulp. Often one type of machine will be used for roughing and another for cleaning.

WASTE TREATMENT PLANTS

Industrial hazardous wastes, i.e., solids, powders, sludges, slurries, liquids, etc., are transportable and may be immediate or future polluters of the land, water, air on both a short- and long-term basis. Wastes must therefore be properly disposed of in accordance with applicable governmental regulations. Waste treatment plants are available to handle industrial hazardous wastes. Waste disposal practices are usually determined by the physical properties of the waste (liquid, sludges, etc.) rather than by its chemical properties. Industrial hazardous waste treatment can be characterized as primary, secondary, and tertiary (or advanced), followed by ultimate disposal of the effluent wastewater and sludge.

The first or primary treatment step generally utilizes gravity differential, i.e., sedimentation or flotation, but can also include chemical treatment such as neutralization and hydrolysis. The next or secondary step is usually biochemical, involving activated sludge, trickling filters, aeration lagoons and the like. Tertiary or advanced treatment consists essentially of mechanical or chemical techniques, such as filtration, carbon adsorption and ion exchange. Ultimate disposal can involve digestion and thickening, followed by incineration, landfarming, or composting.

Primary Treatment

Even before primary treatment processes are applied, preliminary screening and skimming operations may be employed. Solids such as rags, sticks, and trash in the raw wastewater are screened, that is, removed on racks or bar screens placed at the head of the treatment plant. Screenings may be disposed of separately or ground by hammer mills or shredders and added to the wastewater for later removal in sedimentation basins. Heavy inert material or grit such as sand, silt, gravel, and ashes are selectively removed at the head of the wastewater treatment plant by velocity control in simple gravity settling chambers or by buoyant induction in air flotation tanks. In addition, initial floatable materials can be skimmed or collected from sedimentation basins and subsequently digested, dewatered, incinerated, and/or landfarmed.

Neutralization--

Neutralization has wide applicability to waste streams of diverse physical and chemical compositions and would be available in almost any waste treatment facility. Neutralization is a liquid-phase chemical reaction between an acid and a base which produces a neutral solution. It may be carried out in batch or continuous mode and requires reaction tanks, agitators, pumps, and ancillary equipment for handling solids and/or liquids, and storage facilities. The process can be used on aqueous and non-aqueous liquids, slurries, and sludges.

Hydrolysis--
Hydrolysis as a chemical process may be conducted with (1) water alone, (2) aqueous acid, (3) aqueous alkali, (4) alkali fusion with little or no free water, and (5) enzymes as catalysts. Hydrolysis can be conducted in simple systems, e.g., batch-wise in open tanks, or in more complicated equipment, e.g., continuous flow in large towers.

Sedimentation--
Sedimentation is a purely physical process whereby particles suspended in a liquid are allowed to settle by means of gravitational and inertial forces acting on the particles suspended in the liquid and the liquid itself. Sedimentation can be carried out as either a batch or a continuous process, with the latter far more common. The settled particles are either removed from the bulk of the liquid or the liquid is separated from the settled particles by decantation. Sedimentation can be carried out in rudimentary settling ponds, conventional settling basins, or in more advanced clarifiers that are equipped with flocculation zones and tube-like devices that enhance settling. Settling ponds are periodically emptied of particles by mechanical shovels, draglines, or siphons. Sedimentation basins and clarifiers usually employ a built-in solids collection and removal device such as a sludge scraper and draw-off mechanism.

Flotation--
After removal of the majority of free hazardous substance in a separator, remaining free hazardous substance together with colloidal emulsions and suspended solids may be further reduced in a dissolved or induced air flotation unit (DAF or IAF). The basic principle of either DAF or IAF is that air bubbles attach to the suspended hazardous substance or solids causing the particles to float to the surface where they can be skimmed.

The two common types of air flotation dissolving systems differ in the method of air pressurization, based on the waste characteristics. Dispersed or induced air flotation is used for treatment to remove free hazardous substances. Dispersed air flotation induces air into the water or wastewater with mechanical rotors or agitators. Dissolved air flotators are used to treat hazardous substance wastewater streams after initial separation in gravity separation tanks. In dissolved air flotation air is first dissolved in water at elevated pressures and then the pressure is released and the supersaturated water becomes filled with very fine bubbles.

Secondary Treatment

Activated Sludge--
The activated sludge process is the most commonly employed treatment process for the degradation of organics in wastewater. Its chief distinguishing characteristic is that the heterogeneous microbial populations or sludge

biomass are growing in a fluidized state by turbulent mixing. The conventional activated sludge process consists of an aeration tank, a secondary clarifier, and a sludge recycle line. Both raw wastewater and recycled sludge enter the tank at the head end and are aerated for a specified time period. The wastewater and recycled sludge are mixed by the action of diffused or mechanical aeration. During this period, absorption, flocculation, and oxidation of the organic matter take place. The mixed liquor is fed into the settling tank and settled sludge is returned at a fixed rate. Modifications of the conventional process are found in the complete-mix, tapered-aeration and step-aeration processes.

Trickling Filter--
Trickling filters or biological filters are fixed-growth biological systems in which the wastewater is contacted with microbial growths attached to the surfaces of the supporting media. The bacteria growths biochemically oxidize the matter. Proper functioning of trickling filters requires aerobic conditions in the bed and sufficient void space in the packing. The wastewater can be sprayed over a bed of crushed rock or several forms of synthetic media. Installations with synthetic media called biological filters or towers are often about 20 feet in depth rather than the traditional 6 feet for rock-filled filters. The main advantage of synthetic media is the high specific surface with a corresponding high percentage of void volume.

Rotating Biological Contactor--
The rotating biological contactor consists of large-diameter plastic media, mounted on a horizontal shaft. The plastic media is slowly rotated while approximately 40 percent of the surface area is submerged in the wastewater. During rotation, the contactor carries a film of wastewater into the air, which trickles down the surfaces and absorbs oxygen from the air. Micro-organisms adhere to the rotating surfaces and remove organic materials from the film of wastewater.

Aerated Lagoon--
An aerated lagoon is generally a basin of significant depth (usually 6 to 17 feet), in which wastewater is treated on a flow-through basis. Organic waste stabilization is accomplished by a dispersed biological growth system, and oxygenation is usually supplied by means of surface aerators or diffused aeration equipment. The aerated lagoon is analagous to the extended activated sludge process without the sludge return. There are two distinctly different types of aerated lagoons depending on the amount of mixing: the aerobic lagoon and the aerobic-anaerobic or facultative lagoon. The aerobic lagoon is designed with sufficient power input to create suitable turbulence to maintain the solids in suspension. The facultative lagoon is designed with only enough power input to maintain sufficient dissolved oxygen distribution throughout the basin.

Ozonation--
Ozonation consists of treatment with ozone, O_3, an extremely reactive gas and a powerful oxidizing agent. Ozone also has powerful antibacterial and antiviral properties. Ozone must be generated on site immediately prior to application. The produced ozone (ozonized air) is fed through porous

diffusers into an ozonized air/water emulsification tank where the ozone is brought into intimate contact with the oxidizable materials. Provided that excess ozone is not allowed to escape, ozonation systems are effective in hazardous waste applications.

Tertiary Treatment

Filtration--

Filtration is a purely physical process whereby particles suspended in a fluid are separated from it by forcing the fluid through a porous medium. Generally, the following features are common to all filtration systems: 1) a porous filter medium, (2) an induced pressure differential across the filter medium to provide fluid flow through, (3) a mechanical device which contains or supports the filter medium, and (4) a means of removing the entrapped particles from the filter medium. Filter media can be divided into three general classes: (1) a thick barrier composed of granular material, such as sand, coke, coal, or porous ceramics, (2) a thin barrier such as a filter cloth or filter screen, and (3) a thick barrier composed of a disposable material such as powdered diatomaceous earth or waste ash. The pressure differential required for flow can be induced by gravity, positive pressure, or vacuum. Granular media filtration utilizes a bed of granular particles, usually sand, as the filter medium. The bed is typically contained in a basin or tank and is supported by an underdrain system that allows the filtered liquid to be drawn off while retaining the filter medium in place. The underdrain system can consist of metal or plastic strainers located at intervals on the bottom of the filter. A washwater stream is forced through the bed of granular particles in the reverse direction of the original flow to dislodge the filtered solids. The volume of the backwash water stream, however, is only a small fraction (1-4%) of the liquid volume being filtered.

Steam Distillation--

Steam distillation is a physical process for removing water-immiscible volatile organics from a waste stream. It can also be used to recover heat-sensitive, high boiling point, water soluble components from a waste stream. A typical steam distillation system consists of: (1) a feed pump, (2) batch still, (3) condenser, and (4) gravity separator. With a water phase present in steam distillation, the effective boiling point of the immiscible liquid is reduced.

Reverse Osmosis--

The reverse osmosis process derives its name from the fact that the flow of water under the applied driving force is in a direction opposite to that normally observed in osmotic process. The process membrane consists of a surface layer with bound water but little or no capillary water. The surface layer rejects salts because the solution capability of bound water is almost zero. Use of reverse osmosis for industrial waste treatment is growing rapidly. The process is particularly useful for concentrating waste streams containing dissolved organics or inorganics, to facilitate recovery, or to reduce volume.

Wastewater Disposal

The treated (primary, secondary, tertiary) wastewater can be disposed of via one or more of the following routes: (1) discharge into a receiving water, (2) reuse, (3) transport to using site, (4) ocean disposal, (5) seepage into the groundwater, and (6) surface application.

Sludge Treatment

Sludge treatment can consist of: (1) anaerobic digestion, (2) centrifugation, (3) thickening, (4) vacuum filtration, and (5) bed drying (lagooning). Anaerobic digestion involves the biological decomposition of organic and/or inorganic matter in the absence of molecular oxygen. This treatment technique is normally considered a two-stage process. In the first stage no methane is formed and no waste stabilization occurs as the complex organics are changed in form by a group of anaerobic bacteria. In the second stage, true waste stabilization occurs as the gaseous end products, carbon dioxide and methane, are produced.

Centrifuges achieve mechanical dewatering of sludges by application of centrifugal force. The concentration of suspended solids in sludges and slurries by centrifuges depends on settling due to a difference in density between the suspended particle and the liquid medium. Centrifugation technology includes cyclone, disk, scroll and imperforate basket methods of separation.

Gravity thickening or sedimentation for concentration of waste sludges is a common method used to minimize sludge volume. Gravity thickeners may be round or rectangular and sludge is removed from the bottom by rakes or suction piping. Parallel plate devices are used for streams handling dilute solids. Sludge which accumulates in the primary clarifier varies from 2 to 8% solids depending on the operating efficiency of the clarifier and on whether thickening is used. Anaerobic digestion and various dewatering techniques are more easily applied to the sludge from primary clarifiers; however, primary sludge is frequently mixed with secondary sludge prior to treatment.

Two commonly used vacuum filtration apparatuses are the rotary drum filter and the rotary belt filter. A coil filter has been developed that is similar to the belt filter except that metal coil springs are used in place of the filter belt. A vacuum filtration apparatus is a complex system that, in addition to the vacuum filter, includes sludge feeding and conditioning components, vacuum pump and receiver, and filtrate pump.

Drying beds are the most widely used sludge dewatering method and usually consist of open or covered sand beds or paved beds with limited drainage. Sand beds need little operator attention and drying is usually restricted to well digested sludge. The beds usually consist of both a gravel layer 8 to 20 inches deep that extends at least 6 inches above the top of the underdrains and also a covering sand layer between 4 and 12 inches deep. Sand beds can be enclosed by glass or other material where justifiable to protect the drying sludge from rain, control odors, insects, and reduce the drying periods during cold weather. Good ventilation is important to control humidity and optimize the evaporation rate. Lagoon drying is a simple low-cost system of sludge dewatering. In drying lagoons, as in sand beds, the sludge is periodically removed and the land refilled. Lagoon depth should not be more than 24 inches and the soil must be reasonably porous with the bottom of the lagoon at least 18 inches above the maximum groundwater table.

Sludge Disposal

Conventional sludge disposal usually consists of (1) thermal degradation, (2) landfilling, and (3) landfarming. The methods of thermal degradation include: co-firing in boilers, incineration, and pyrolysis. Such methods for treatment of waste produce secondary emissions and, therefore, are not considered to be ultimate methods of disposal. The secondary emissions must undergo additional treatment before being safely discharged into the receiving environment. Such secondary discharges are closely controlled by a host of federal, as well as state and local, legislation.

Landfilling and landfarming, at this time, are the most common and widely used methods for disposal of hazardous wastes. The EPA controls hazardous waste disposal in landfills under RCRA regulations. Key design features of a landfill facility include cells used to separate incompatible wastes and also runoff control systems. The cells may be physically separate areas, trenches, or pits with liners to provide isolation of wastes to from adjacent cells. Runoff from active portions of a landfill must be collected and treated. Landfill disposal must be at a permitted landfill and the waste must be accompanied by a Universal Waste Manifest.

Landfarming has been used for many years as an effective method for disposal of certain sludges. The landfarm site is relatively flat and the groundwater should be at least several feet below the surface. A loam-type soil is ideal. Runoff is controlled by berms around the perimeter and by use of a gate valve to control runoff which must be routed to a water treatment system. The maximum rate of degradation is obtained when the initial substance-in-soil loading is maintained between 5 and 10 percent. For fluid sludges, application is by a mud pump and irrigation pipe or fire hose. Viscous or semisolid sludges can be hauled by dump truck and spread with a bulldozer or grader.

With the enactment of the RCRA reauthorization bill (H.R. 2867), conventional and inexpensive methods for ultimate disposal of hazardous wastes will cease to exist. Less conventional methods will become more prevalent. Two such methods for disposal of sludge are encapsulation or solidification and chemical detoxification.

Other Noyes Publications

TREATMENT OF HAZARDOUS WASTE LEACHATE

Unit Operations and Costs

by

J.L. McArdle, M.M. Arozarena, W.E. Gallagher

PEI Associates, Inc.

Pollution Technology Review No. 151

This book describes 20 technologies that have been employed to treat leachates from hazardous waste sites. The USEPA's hazardous waste site cleanup program, referred to as Superfund, was authorized and established in 1980 by the enactment of CERCLA. This legislation allows the federal government (and cooperating state governments) to respond directly to releases and the threat of releases of hazardous substances and pollutants or contaminants that could endanger public health or welfare or the environment. This book provides guidance in the treatment of hazardous waste leachate, one type of release covered by CERCLA.

The 20 unit operations presented are reviewed for their applicability to the treatment of hazardous waste leachate. They are classified into the following four categories: pretreatment operations, physical/chemical treatment processes, biological treatment, and post-treatment operations. Typical treatment process trains are also presented for leachate containing organic and/or inorganic contaminants.

Factors affecting leachate generation and its compositions are addressed. Also discussed is the management of residuals—sludges, air emissions, concentrated liquid waste streams, spent carbon—generated by the various treatment techniques.

CONTENTS

ISBN 0-8155-1160-4 (1988) **111 pages**

Other Noyes Publications

UNDERGROUND TANK LEAK DETECTION METHODS

by

Shahzad Niaki and John A. Broscious
IT Corporation

Pollution Technology Review No. 139

This state-of-the-art review details available and developing methods for detecting leaks in underground storage tanks. Thirty-six volumetric, non-volumetric, inventory monitoring, and leak effects monitoring detection methods are described for underground tanks used primarily for liquid hydrocarbons and products. The book provides general engineering comments on the methods and discusses variables which may affect their accuracy.

In recent years, the increase in leaks from underground fuel storage tanks has had a significant adverse impact on the United States. Current estimates are that there are from 1.5 to 3.5 million of these storage tanks in the U.S. As many as 100,000 may have some leakage already, and it is anticipated that up to 350,000 may develop leaks within the next five years.

Corrosion is one of the primary causes of leakage. Product loss from leaking tanks may adversely affect the environment, endanger lives, reduce income, and require expensive cleanup procedures. To prevent, reduce, or avoid such problems, accurate methods must be used to determine whether or not an underground tank is leaking. This book identifies existing and developing techniques for leak detection. In so doing, each method, its capabilities, its claimed precision and accuracy, are reviewed.

The condensed table of contents given below lists **chapter titles and selected subtitles.**

ISBN 0-8155-1117-5 (1987) **123 pages**